AF870373

ELECTRONIC STRUCTURE OF ATOMS, MOLECULES AND SOLIDS

ELECTRONIC STRUCTURE OF ATOMS, MOLECULES AND SOLIDS

BRAZILIAN SCHOOL ON ELECTRONIC STRUCTURE

Editors

Sylvio Canuto
Departamento de Física,
Universidade Federal de Pernambuco
50.739 Recife, PE, Brazil

José D' Albuquerque e Castro
Instituto de Física, Universidade Federal Fluminense
24.020 Rio de Janeiro, RJ, Brazil

Fernando J. Paixão
Instituto de Física
Universidade Estadual de Campinas
13.081 Campinas, SP, Brazil

World Scientific
Singapore • New Jersey • London • Hong Kong

Published by

World Scientific Publishing Co Pte Ltd
5 Toh Tuck Link, Singapore 596224
USA office: 27 Warren Street, Suite 401-402, Hackensack, NJ 07601
UK office: 57 Shelton Street, Covent Garden, London WC2H 9HE

British Library Cataloguing-in-Publication Data
A catalogue record for this book is available from the British Library.

ELECTRONIC STRUCTURE OF ATOMS, MOLECULES AND SOLIDS

Copyright © 1990 by World Scientific Publishing Co Pte Ltd

All rights reserved. This book, or parts thereof, may not be reproduced in any form or by any means, electronic or mechanical, including photocopying, recording or any information storage and retrieval system now known or to be invented, without written permission from the publisher.

For photocopying of material in this volume, please pay a copying fee through the Copyright Clearance Center, Inc., 222 Rosewood Drive, Danvers, MA 01923, USA. In this case permission to photocopy is not required from the publisher.

ISBN-13 9789810201197
ISBN-10 9810201192

PREFACE

There has been a great deal of development in some areas of electronic structure of matter with a high degree of specialization. As a consequence it is becoming more difficult to follow the developments in neighbouring areas. This proceedings volume tries to fill the gap and collect in one single volume topics such as superlattice structure, polymers, density functional theory, molecular photoionization, impurities in semiconductors, to mention a few. The contributions are selected to emphasize most of the active areas and aimed at graduate students in the field but should also be useful, we hope, to the experienced professional.

This II Brazilian School on Electronic Structure was held in July 1989. It took place in a monastery built in 1535, located in the historic site of Olinda. It has been promoted by the Brazilian Physical Society. We acknowledge the financial support from several Brazilian Agencies (FINEP, CNPq, CAPES, FAPESP and FAPERJ), as well as the British Council.

The Editors

CONTENTS

Molecular photoionisation cross sections by moment theory. An introduction

Florian Müller-Plathe
SERC Daresbury Laboratory, Warrington WA4 4AD, U.K.

and

Geerd H. F. Diercksen
Max-Planck-Institut für Physik und Astrophysik, Institut für Astrophysik
Karl-Schwarzschild-Straße 1, D-8046 Garching bei München, F.R.G.

1. Introduction

Moment theory is known in mathematics since the turn of the century. It goes back to the work of Chebyshev, Markov, Stieltjes and Hamburger, to name but a few (see refs. [1, 2]). It borders on many other fields in mathematics such as: orthogonal polynomials [3, 4], continued fractions and Padé approximants [5], and Gaussian quadrature [6]. However, its potential in physics has only been recognised in the late 1960's when it was applied in the context of statistical mechanics of solids [7-12]. It was then used in atomic and molecular physics in determining electrical response properties like dynamic polarisabilities, refraction indexes, Rayleigh-scattering cross sections, Verdet constants for the molecular Faraday effect, and C_6 coefficients for the dispersion interactions [13-15]. After application to the calculation of photoionisation cross sections, which is the subject of the present treatise, moment theory has also been used in a variety of other applications: electron-impact ionisation cross sections [16], resonance widths [17], shake-off ionisation [18], and calculation of integrals over the density of states [19]. More recently, the method has been extended to provide approximate continuum wavefunctions within certain models [20-22].

It is our intention to provide a technical review of the algorithms. Basic aspects of the oscillator strength and its distribution are dealt with in section 2. The moment theory part is presented in section 3 with the photoionisation problem in mind. However, the formulations should be easily adaptable to other applications as well. Special attention has to be paid to the computational methods used to calculate the transition energies and oscillator strength which enter the moment-theoretical formalism. Many of the erroneous results of earlier moment-theoretical calculations of photoionisation cross sections are really caused by using inappropriate methods for the calculation of these transition parameters. In our opinion, linear-response methods are particularly suitable for this purpose, and in section 4 their relevant features are described. In section 5 an attempt is made to sketch the guidelines for devising basis sets for these calculations, although the experience is fairly limited as yet. In section 6 it is tried to establish the range of applicability of the method and to describe some of the typical pitfalls.

In this review no applications are discussed. Instead, references to the original articles are provided.

2. Fundamentals

The probability for a photon of energy ω to be absorbed by a molecule is called the cross section $\sigma(\omega)$. It is proportional to the (differential) oscillator strength $f(\omega)$

$$\sigma(\omega)/\mathrm{Mb} = 2\pi^2\alpha a_0^2 \times 10^{18} f(\omega) \tag{2.1}$$

where 1 Megabarn (Mb) $= 10^{-18}$ cm^2, α is the fine structure constant, and a_0 denotes the Bohr radius in centimetres. The oscillator strength is dimensionless.

Throughout this paper we restrict ourselves to absorption processes in which one photon is absorbed and one electron is excited or ejected (ionised). However, the term "ionisation" is to be understood in a general sense as a process during which an electron is removed from a molecular system into the continuum. This includes processes where a neutral molecule is transformed into a singly-charged positive molecular ion, which is usually understood as genuine ionisation, and processes where one electron is ejected from a charged particle, in which case the more precise term is "photodetachment".

The oscillator strength f_i for a photoexcitation from the ground state $|0\rangle$ to some excited state $|i\rangle$ is given by the golden-rule expressions (in atomic units au)

$$f_i = \frac{2}{3}\omega_i |\langle 0| \sum_j \mathbf{r}_j |i\rangle|^2 \tag{2.2}$$

$$f_i = \frac{2}{3}\frac{1}{\omega_i} |\langle 0| \sum_j \mathbf{p}_j |i\rangle|^2 \tag{2.3}$$

$$f_i = i\frac{2}{3} \langle 0| \sum_j \mathbf{r}_j |i\rangle\langle i| \sum_j \mathbf{p}_j |0\rangle \tag{2.4}$$

where ω_i is the energy difference between the excited state and the ground state, $\mathbf{r}_j$ and $\mathbf{p}_j$ are the position and momentum operators for electron j, respectively, and the sums run over all electrons in the molecule. These formulae involve already a number of approximations, the most important of which is the dipole approximation which holds as long as the wavelength of the incident photon is much larger than the dimensions of the molecule. The above three forms are commonly denoted as the dipole-length (DL), dipole-velocity (DV) and mixed dipole-length dipole-velocity (DLDV) formulation of the oscillator strength. It has to be emphasised that these formulae are exact within the dipole approximation only for both states being exact solutions of the electronic Schrödinger equation, but not for approximate wave functions as usually encountered in quantum chemistry (with a few exceptions [23]). Similarly, the DL, DV and DLDV forms of the oscillator strength are exactly equivalent to each other only for true eigenstates and for certain well-defined approximate eigenfunctions, but not for approximate eigenfunctions in general.

In the case of photoionisation the discrete oscillator strengths f_i are replaced by a continuous oscillator strength function $f(\omega)$ which can be calculated in an analogous way in the three formulations

$$f(\omega) = \frac{2}{3}\omega |\langle 0| \sum_j \mathbf{r}_j |\phi(\omega)\rangle|^2 \tag{2.5}$$

$$f(\omega) = \frac{2}{3}\frac{1}{\omega} |\langle 0| \sum_j \mathbf{p}_j |\phi(\omega)\rangle|^2 \tag{2.6}$$

$$f(\omega) = i\frac{2}{3} \langle 0| \sum_j \mathbf{r}_j |\phi(\omega)\rangle\langle \phi(\omega)| \sum_j \mathbf{p}_j |0\rangle \tag{2.7}$$

where $\phi(\omega)$ is a continuum solution of the electronic Schrödinger equation with an energy ω above the ground state. The wavefunctions $\phi(\omega)$ form a continuous spectrum (hence their name), whereas the bound state wavefunctions form a discrete, denumerable set.

The continuum wavefunctions $\phi(\omega)$ are in some respects very different from discrete bound state wavefunctions. Continuum wavefunctions tend to oscillate between two non-zero limiting values, whilst bound-state wavefunctions asymptotically approach zero at infinity. For atoms and molecules they become plane waves $\exp(i\mathbf{k}\cdot\mathbf{r})$ at large distances. Therefore, unlike bound-state wavefunctions, they are not square-integrable, i.e. the integral $\int \phi^*(\omega)\phi(\omega)d\mathbf{r}$ is infinite.

The moments of the molecular oscillator strength $s(k)$

$$s(k) = \sum_{\text{discr.}} \omega_i^k f_i + \int_{\text{cont.}} \omega^k f(\omega) d\omega \tag{2.8}$$

that include excitations to discrete states as well as ionisation into the continuum play a vital role in the method described in this article and therefore merit a closer look. The oscillator strength moments are, explictly or implicitly, the input to moment-theoretical calculations. Accurate oscillator strength moments are required for calculating reliable photoionisation cross sections. In order to check their accuracy one would like to compare the moments directly to experimental information if possible.

Fortunately, several of the total oscillator strength moments $s(k)$ are related by sum rules to physical observables or to expectation values of certain operators over the ground state wavefunction $|0\rangle$, [24-26]

$$s(-2) = \bar{\alpha} \tag{2.9}$$

$$s(-1) = \frac{2}{3}\langle 0|(\sum_{i=1}^{N} \mathbf{r}_i)^2|0\rangle \tag{2.10}$$

$$s(0) = N \tag{2.11}$$

$$s(1) = \frac{2}{3}\langle 0|(\sum_{i=1}^{N} \mathbf{p}_i)^2|0\rangle \tag{2.12}$$

$$s(2) = \frac{4}{3}\pi N\langle 0|(\sum_{i=1}^{N} \delta(\mathbf{r}_i))^2|0\rangle \tag{2.13}$$

Here, $\bar{\alpha}$ denotes the average static electric dipole polarisability, N the number of electrons in the molecule, $\mathbf{r}_i$ and $\mathbf{p}_i$ the position and momentum operators of electron

i, and all quantities are in atomic units. Again, the sum rules apply to the exact wavefunctions and to certain approximations, provided that the basis-set error is negligible. The sum rules $s(-2)$ and $s(0)$ are particularly important in the present context. The dipole polarisability $s(-2)$ is the highest negative moment that can be measured conveniently. If the polarisability is not reproduced, it is very likely that the higher negative moments are incorrect as well. The Thomas-Reiche-Kuhn (TRK) sum rule $s(0)$ normalises the oscillator strength distribution, so that it must be obeyed in order to avoid global over- or underestimation of the cross section.

Closely related to the normal sum rules are the logarithmic sum rules $l(k)$

$$l(k) = \sum_{\text{discr.}} \omega_i^k \ln \omega_i f_i \; + \int_{\text{cont.}} \omega^k \ln \omega f(\omega) d\omega \tag{2.14}$$

They are used to define the so-called mean excitation energies $I(k)$

$$I(k) = \exp\left(\frac{l(k)}{s(k)}\right) \tag{2.15}$$

The $I(k)$ appear in the Bethe treatment of charged particle impact on matter [27-29]: $I(-1)$ is related to the cross section for electron excitation/ionisation, $I(0)$ to the energy deposition (stopping power), and $I(1)$ to its mean fluctuation or straggling. The Lamb shift of electronic energy levels is governed by $I(2)$ [26]. The mean excitation energies can also be used to probe the quality of the oscillator strength distribution.

3. Algorithms in moment theory

3.1. General remarks

The moment-theoretical approach to photoionisation cross section calculations relies on the fact that the moments of the continuum oscillator strength $S(k)$

$$S(k) = \int_{\omega_T}^{\infty} \omega^k f(\omega) d\omega \tag{3.1.1}$$

where ω_T is the ionisation threshold energy, can be well approximated by a corresponding sum over discrete pseudostates obtained from a quantum-chemical calculation in a discrete basis set

$$S(k) \approx \sum_{i=1}^{N} \bar{\omega}_i^k \bar{f}_i \tag{3.1.2}$$

where the overbar denotes energies and oscillator strength for excitations to pseudostates that fall into the ionisation continuum. However, since the pseudostates arise from a calculation performed in a square-integrable basis set (commonly referred to as an L^2 basis) they have the wrong boundary conditions, viz the boundary conditions for bound states. These boundary conditions cause them to form a discrete, rather than a continuous spectrum. Therefore, in general no physical reality whatsoever can be ascribed to the pseudostates, and also the excitation energies and oscillator strength for transitions to these pseudostates are physically meaningless. Moreover, they depend strongly on the basis set and the quantum-chemical technique used in the calculation. However, the moments obtained from them indeed approximate the true oscillator strength moments, provided that certain conditions are met. The physical rationale behind the moment-theoretical approach is based on the following observation: the pseudostate wavefunction approaches closely the behaviour of the true continuum wavefunction in the molecular core region, if the basis set is sufficiently flexible. This region is most important for the absorption process because in the initial bound state the electrons are practically confined to this region. A more thorough treatment is given by Reinhardt [30].

3.2. The moment problem

The moments $S(k)$ of a nondecreasing distribution function $F(\omega)$ over the integration interval $[a,b]$ are defined as

$$S(k) = \int_a^b \omega^k dF(\omega) \qquad k = 0,1,\ldots \tag{3.2.1}$$

For our purpose a less general definition of the moments is sufficient. It follows from the assumption

$$dF(\omega) = \frac{dF(\omega)}{d\omega} d\omega = f(\omega) d\omega \tag{3.2.2}$$

i.e.

$$f(\omega) = \frac{dF(\omega)}{d\omega} \qquad F(\omega) = \int_a^\omega f(\omega') d\omega' \tag{3.2.3}$$

Hence

$$S(k) = \int_a^b \omega^k f(\omega) d\omega \qquad k = 0,1,\ldots \tag{3.2.4}$$

The function $f(\omega)$ is restricted to be nonnegative in $[a,b]$. In statistics it is referred to as the density function and in Gaussian quadrature as the weight function. In the photoionisation case $f(\omega)$ is the differential continuum oscillator strength which is sought, whereas $F(\omega)$ is the cumulative oscillator strength.

With these definitions we can state the moment problem [1].

Moment problem: Given the moments $S(k)$, with $k = 1, 2, 3, \ldots$, determine the weight function $f(\omega)$ or the distribution function $F(\omega)$.

In practical applications usually only the first few moments are available, apart from a few cases where the moments can be calculated analytically. Therefore, a somewhat different problem has to be solved.

Reduced moment problem: Given a finite set of moments $S(k)$, with $k = 0, 1, \ldots$, $2r - 1$, determine a function $f(\omega)$ so that it reproduces all available moments in the interval $[a, b]$

$$S(k) = \int_a^b \omega^k f(\omega) d\omega \qquad k = 0, 1, \ldots, 2r - 1 \tag{3.2.5}$$

Note that a finite set of moments does not determine $f(\omega)$ completely. In general, there are infinitely many weight functions $f(\omega)$ which reproduce the first few moments. This is in marked contrast to the full moment problem, where a complete set of moments (which may be infinitely many) fully determines $f(\omega)$.

In order to solve the reduced moment problem a procedure originally proposed by Chebyshev is used. It selects a unique solution $F(\omega)$ from the infinite number of solutions. The Chebyshev algorithm finds approximate distributions $F^{(n)}(\omega)$ and densities $f^{(n)}(\omega)$ of different orders n up to the maximum order r, depending on the number of moments $2r$. For these approximate distributions there exist useful theorems, the so-called *Chebyshev inequalities* [1]

$$F^{(n)}(\omega_i - 0) \leq F^{(n+1)}(\omega_i - 0) \leq F(\omega_i) \leq F^{(n+1)}(\omega_i + 0) \leq F^{(n)}(\omega_i + 0) \tag{3.2.6}$$

where the ω_i are certain abscissae, which will be discussed below, and $\omega_i - 0$ and $\omega_i + 0$ denote limits from below and above, respectively. There are two important implications of the Chebyshev inequalities:

- at certain points ω_i the approximate distributions $F^{(n)}(\omega)$ converge to the "true" (infinite order) distribution $F(\omega)$ with increasing n.
- the lower and the upper limit of the approximate distribution $F^{(n)}(\omega)$ bracket the "true" value of the distribution $F(\omega)$ at the ω_i. In other words $F^{(n)}(\omega_i - 0)$ and $F^{(n)}(\omega_i + 0)$ provide the lower and the upper bound to $F(\omega_i)$, respectively.

The algorithm for finding $f^{(n)}(\omega)$, commonly known as "Stieltjes imaging", is described in the following and is implemented in the program SCAMPI ("Stieltjes-Chebyshev Analysis of Molecular Photoionisation") [31]. The procedure involves several steps which are dealt with in detail in the following subsections (3.3.-3.5.). The procedure is not limited to photoionisation studies but is quite general and any nonnegative density whose moments are finite and known can be "imaged". In the context of photoionisation negative oscillator strength moments ($k \leq 0$) have to be used because positive moments with $k > 2$ diverge [26]. Some applications, however, may start from positive moments. For these cases it might be necessary to reformulate the algorithms accordingly. We briefly outline an alternative approach for solving the reduced moment problem, namely the expansion of the density into orthogonal polynomials in subsection 3.6.

3.3. Generation of orthogonal polynomials

The moments $S(-k)$, $k = 0, 1, \ldots, 2r - 1$ define a sequence of orthogonal polynomials $Q_n(1/\omega) = \sum_{i=0}^{n} Q_n^i (\frac{1}{\omega})^i$ of degree 0 to r [5] which are orthogonal in the interval $[a, b]$ (in the photoionisation case $[\omega_T, \infty]$) with respect to the unknown weight function $f(\omega)$

$$\int_a^b Q_n(1/\omega) Q_m(1/\omega) f(\omega) d\omega = N_n \delta_{nm} \tag{3.3.1}$$

The normalisation $N_n = \int_a^b [Q_n(1/\omega)]^2 f(\omega) d\omega$ is chosen so that the coefficient of the highest power $(1/\omega)^n$ in the polynomial $Q_n(1/\omega)$ is unity. These monic polynomials satisfy the three-term recurrence relations

$$Q_n(1/\omega) = (1/\omega - a_n) Q_{n-1}(1/\omega) - b_{n-1} Q_{n-2}(1/\omega) \tag{3.3.2}$$

with the special cases

$$Q_0(1/\omega) = 1 \quad \text{and} \quad Q_1(1/\omega) = 1/\omega - a_1 \tag{3.3.3}$$

The recurrence coefficients a_n and b_{n-1} are given by [6]

$$a_n = \frac{\int_a^b (1/\omega)[Q_{n-1}(1/\omega)]^2 f(\omega) d\omega}{\int_a^b [Q_{n-1}(1/\omega)]^2 f(\omega) d\omega} \tag{3.3.4}$$

$$b_{n-1} = \frac{\int_a^b [Q_{n-1}(1/\omega)]^2 f(\omega) d\omega}{\int_a^b [Q_{n-2}(1/\omega)]^2 f(\omega) d\omega} \tag{3.3.5}$$

By substituting in the formulae (3.3.4) and (3.3.5) the explicite form of the polynomials $Q_{n-1}(1/\omega)$ and $Q_{n-2}(1/\omega)$ and by exchanging the order of integration and summation the numerators and denominators can be expressed as sums of the known moments $S(-k)$ each weighted with a coefficient arising from the multiplication of the polynomials. The highest power of $(1/\omega)$ occurring in the integrands of eqns. (3.3.4) and (3.3.5) is $(1/\omega)^{2n-1}$. Hence, if the moments from $S(0)$ through $S(-(2r-1))$ are available all polynomials of degree up to r can be determined.

The expressions for a_n and b_{n-1} can be cast into forms more convenient for computation. Making use of the equation

$$\int [Q_{n-1}(1/\omega)]^2 f(\omega) d\omega = b_0 b_1 \cdots b_{n-1} \tag{3.3.6}$$

(the integration limits are omitted for brevity) we obtain the alternative equations [32]

$$a_n = \frac{1}{b_0 b_1 \cdots b_{n-1}} \int (1/\omega) [Q_{n-1}(1/\omega)]^2 f(\omega) d\omega \tag{3.3.7}$$

$$b_{n-1} = \frac{1}{b_0 b_1 \cdots b_{n-2}} \int [Q_{n-1}(1/\omega)]^2 f(\omega) d\omega \tag{3.3.8}$$

An equivalent form is given by Wall [5]

$$a_n = \frac{1}{b_0 b_1 \cdots b_{n-1}} \int (1/\omega)^n Q_{n-1}(1/\omega) f(\omega) d\omega - \sum_{l=1}^{n-1} a_l \tag{3.3.9}$$

$$b_{n-1} = \frac{1}{b_0 b_1 \cdots b_{n-2}} \int (1/\omega)^{n-1} Q_{n-1}(1/\omega) f(\omega) d\omega \tag{3.3.10}$$

The equivalence of the two sets of equations is shown using a fundamental property of orthogonal polynomials: the orthogonal polynomial $Q_n(1/\omega)$ of degree n is orthogonal to any arbitrary (not necessarily orthogonal) polynomial in $1/\omega$ whose degree is lower than n. Hence the two equations for b_{n-1} (3.3.8 and 3.3.10) are equivalent because $Q_{n-1}(1/\omega)$ is orthogonal to the simple powers $(1/\omega)^m$ with $m < n-1$, which are polynomials with only one term.

The proof of the equivalence of the two equations for the a_n (3.3.7 and 3.3.9) is somewhat longer. Applying the same orthogonality consideration as before to eqn. (3.3.7) results in

$$a_n = \frac{1}{b_0 b_1 \cdots b_{n-1}} \int (1/\omega)^n Q_{n-1}(1/\omega) f(\omega) d\omega$$
$$+ Q_{n-1}^{n-2} \frac{1}{b_0 b_1 \cdots b_{n-1}} \int (1/\omega)^{n-1} Q_{n-1}(1/\omega) f(\omega) d\omega \tag{3.3.11}$$

where Q_n^m denotes the coefficient of the mth power of $(1/\omega)$ in the orthogonal polynomial of degree n. Both the integral and the denominator in the second term equal N_{n-1}. This leaves

$$a_n = \frac{1}{b_0 b_1 \cdots b_{n-1}} \int (1/\omega)^n Q_{n-1}(1/\omega) f(\omega) d\omega + Q_{n-1}^{n-2} \tag{3.3.12}$$

where Q_{n-1}^{n-2} can be obtained from the three-term recurrence relation for orthogonal polynomials (3.3.2)

$$Q_{n-1}^{n-2} = Q_{n-2}^{n-3} - a_{n-1} Q_{n-2}^{n-2} \tag{3.3.13}$$

By the normalisation convention the orthogonal polynomials are monic, i.e. $Q_{n-2}^{n-2} = 1$. Iterating this equation yields

$$Q_{n-1}^{n-2} = Q_{n-3}^{n-4} - (a_{n-1} + a_{n-2}) \tag{3.3.14}$$

$$\vdots$$

$$Q_{n-1}^{n-2} = Q_1^0 - (a_{n-1} + a_{n-2} + \cdots + a_2) \tag{3.3.15}$$

Since $Q_1^0 = -a_1$ (see eqn. 3.3.3) we have

$$Q_{n-1}^{n-2} = -\sum_{l=1}^{n-1} a_l \tag{3.3.16}$$

Inserting eqn. (3.3.16) into eqn. (3.3.12) completes the proof of the equivalence of the two equations for a_n.

In the application to photoionisation cross sections – and in similar applications where the (approximate) moments are given as appropriate sums – both sets of equations for the recurrence coefficients (eqns. 3.3.7–3.3.10) may be recast into a form which proceeds directly from the transition energies and oscillator strengths. They bypass the intermediate calculation of the moments and are computationally more convenient and allegedly more robust numerically [32]. By expanding the integral

$\int (1/\omega)^n Q_{n-1}(1/\omega) f(\omega) d\omega$ of eqn. (3.3.9) into a sum over moments, substituting the formulae for the approximate moments from eqn. (3.1.2), swapping the summations and regrouping we obtain

$$\int (1/\omega)^n Q_{n-1}(1/\omega) f(\omega) d\omega = \sum_{i=1}^{N} (1/\bar{\omega}_i)^n Q_{n-1}(1/\bar{\omega}_i) \bar{f}_i \tag{3.3.17}$$

and similar expressions for the other integrals in eqns. (3.3.7–3.3.10). Hence, we have for the recurrence coefficients

$$a_n = \frac{1}{b_0 b_1 \ldots b_{n-1}} \sum_{i=1}^{N} (1/\bar{\omega}_i)[Q_{n-1}(1/\bar{\omega}_i)]^2 \bar{f}_i \tag{3.3.18}$$

$$b_{n-1} = \frac{1}{b_0 b_1 \ldots b_{n-2}} \sum_{i=1}^{N} [Q_{n-1}(1/\bar{\omega}_i)]^2 \bar{f}_i \tag{3.3.19}$$

or in the second formulation

$$a_n = \frac{1}{b_0 b_1 \ldots b_{n-1}} \sum_{i=1}^{N} (1/\bar{\omega}_i)^n Q_{n-1}(1/\bar{\omega}_i) \bar{f}_i - \sum_{l=1}^{n-1} a_l \tag{3.3.20}$$

$$b_{n-1} = \frac{1}{b_0 b_1 \ldots b_{n-2}} \sum_{i=1}^{N} (1/\bar{\omega}_i)^{n-1} Q_{n-1}(1/\bar{\omega}_i) \bar{f}_i \tag{3.3.21}$$

The three-term recurrence relation (eqn. 3.3.2) takes the form

$$Q_n(1/\bar{\omega}_i) = (1/\bar{\omega}_i - a_n) Q_{n-1}(1/\bar{\omega}_i) - b_{n-1} Q_{n-2}(1/\bar{\omega}_i) \tag{3.3.22}$$

with the special cases

$$Q_0(1/\bar{\omega}_i) = 1 \quad \text{and} \quad Q_1(1/\bar{\omega}_i) = (1/\bar{\omega}_i) - a_1 \tag{3.3.23}$$

All these recurrence algorithms are numerically unstable for higher values of n, since taking the difference of large numbers tends to accumulate round-off errors very quickly. We have found that the errors become unacceptably large for $n \approx 15$ if 64-bit precision arithmetic is used, and for $n \approx 25$ if 128-bit precision is employed. However, because of the inaccuracy of the approximate higher moments one would not go to high-order orthogonal polynomials anyway and therefore in most cases 64-bit precision is good enough. The problem of stability, though, becomes serious in special cases where the availability of analytical moments allows one to calculate high-order approximations.

3.4. Generalised Gaussian quadrature abscissae and weights

The Gaussian quadrature [6] is the numerical integration rule which for a given number of meshpoints n leads to the best approximation of an integral over $[a, b]$ of a general function $g(\omega)$ weighted with a known weight function $f(\omega)$ by a summation

$$\int_a^b g(\omega)f(\omega)d\omega \approx \sum_{i=1}^{n} f_i g(\omega_i) \tag{3.4.1}$$

The determination of the quadrature abscissae ω_i and the corresponding quadrature weights f_i for the generalised Gaussian quadrature of order n is a necessary step in the solution of the moment problem. These parameters may be obtained from the polynomials orthogonal in $[a, b]$ with respect to the weight function $f(\omega)$. Although no numerical integration are to be performed during the solution of the moment problem we use the terminology of Gaussian quadrature in this subsection.

The abscissae $(1/\omega_i)$ for the n-point quadrature are the roots (zeros) of the orthogonal polynomial $Q_n(\omega)$

$$Q_n(1/\omega_i) = 0 \qquad i = 1, 2, \ldots, n \tag{3.4.2}$$

whereas the weights f_i are obtained from the polynomials of degree lower than n [6]

$$f_i = \left[\sum_{m=0}^{n-1} \frac{[Q_m(1/\omega_i)]^2}{N_m}\right]^{-1} \tag{3.4.3}$$

There are alternative formulae to be found in the literature as well [1, 6, 33]

$$f_i = -\frac{1}{Q_{n+1}(1/\omega_i)Q'_n(1/\omega_i)} \tag{3.4.4}$$

$$f_i = \frac{P_{n-1}(1/\omega_i)}{Q'_n(1/\omega_i)} \tag{3.4.5}$$

where $P_{n-1}(1/\omega)$ is the so called numerator polynomial corresponding to $Q_n(1/\omega_i)$ which can be obtained from a similar recurrence relation

$$P_{n-1}(1/\omega) = (1/\omega - a_n)P_{n-2}(1/\omega) - b_{n-1}P_{n-3}(1/\omega) \tag{3.4.6}$$

with

$$P_{-1}(1/\omega) = 0 \quad \text{and} \quad P_0(1/\omega) = b_0 \tag{3.4.7}$$

For finding the roots of $Q_n(1/\omega)$ (eqn. 3.4.2) any numerical root-search procedure can be used (see for example reference [34]). In fact, very fast Newton-Raphson methods have been devised for the roots of some of the classical orthogonal polynomials [35-36]. Their performance, however, depends on good initial guesses for the roots, which are unavailable in the moment problem. It has therefore become customary to use a more general approach [6, 11, 37], which is sketched in the following.

A set of orthonormal polynomials $\{R_n(1/\omega)\}$

$$\int_a^b R_n(1/\omega)R_m(1/\omega)f(\omega)d\omega = \delta_{mn} \tag{3.4.8}$$

is related to $\{Q_n(1/\omega)\}$ by

$$Q_n(1/\omega) = (-1)^n N_n^{1/2} R_n(1/\omega) \tag{3.4.9}$$

Inserting the $\{R_n(1/\omega)\}$ into the recurrence relation (3.3.2) yields after rearrangement

$$(1/\omega)R_{n-1}(1/\omega) = -\sqrt{b_n}R_n(1/\omega) + a_n R_{n-1}(1/\omega) - \sqrt{b_{n-1}}R_{n-2}(1/\omega) \tag{3.4.10}$$

with the special case

$$(1/\omega)R_0(1/\omega) = -\sqrt{b_1}R_1(1/\omega) + a_1 R_0(1/\omega) \tag{3.4.11}$$

The recurrence relations for all polynomials up to degree n may be written as a matrix equation

$$\begin{pmatrix} a_1 & -\sqrt{b_1} & & & & \\ -\sqrt{b_1} & a_2 & -\sqrt{b_2} & & & \\ & -\sqrt{b_2} & a_3 & -\sqrt{b_3} & & \\ & & \ddots & \ddots & \ddots & \\ & & & -\sqrt{b_{n-2}} & a_{n-1} & -\sqrt{b_{n-1}} \\ & & & & -\sqrt{b_{n-1}} & a_n \end{pmatrix} \begin{pmatrix} R_0(1/\omega) \\ R_1(1/\omega) \\ R_2(1/\omega) \\ \vdots \\ R_{n-2}(1/\omega) \\ R_{n-1}(1/\omega) \end{pmatrix}$$

$$= (1/\omega) \begin{pmatrix} R_0(1/\omega) \\ R_1(1/\omega) \\ R_2(1/\omega) \\ \vdots \\ R_{n-2}(1/\omega) \\ R_{n-1}(1/\omega) \end{pmatrix} - \begin{pmatrix} 0 \\ 0 \\ 0 \\ \vdots \\ 0 \\ -\sqrt{b_n}R_n(1/\omega) \end{pmatrix} \tag{3.4.12}$$

For $(1/\omega_i)$ being a root of $R_n(1/\omega)$ and hence a root of $Q_n(1/\omega)$ the last vector vanishes completely and the matrix equation becomes an eigenvalue equation

$$\begin{pmatrix} a_1 & -\sqrt{b_1} & & & & & \\ -\sqrt{b_1} & a_2 & -\sqrt{b_2} & & & & \\ & -\sqrt{b_2} & a_3 & -\sqrt{b_3} & & & \\ & & \ddots & \ddots & \ddots & & \\ & & & -\sqrt{b_{n-2}} & a_{n-1} & -\sqrt{b_{n-1}} \\ & & & & -\sqrt{b_{n-1}} & a_n \end{pmatrix} \begin{pmatrix} R_0(1/\omega_i) \\ R_1(1/\omega_i) \\ R_2(1/\omega_i) \\ \vdots \\ R_{n-2}(1/\omega_i) \\ R_{n-1}(1/\omega_i) \end{pmatrix}$$

$$= (1/\omega_i) \begin{pmatrix} R_0(1/\omega_i) \\ R_1(1/\omega_i) \\ R_2(1/\omega_i) \\ \vdots \\ R_{n-2}(1/\omega_i) \\ R_{n-1}(1/\omega_i) \end{pmatrix} \tag{3.4.13}$$

The quadrature abscissae $(1/\omega_i)$ may thus be obtained as the eigenvalues of the real symmetric tridiagonal matrix formed by the recurrence coefficients. For large n this diagonalisation is the most time-consuming part of the calculation, but it can be performed reliably and efficiently by standard software [38].

Numerical diagonalisation procedures usually yield eigenvectors $\mathbf{u}_i$ of unit length so that the quadrature weights can not be extracted from them directly. However, by virtue of eqn. (3.4.3) the following equations hold for the weights

$$1 = f_i \sum_{m=0}^{n-1} R_m^2(1/\omega_i) = \mathbf{u}_i \cdot \mathbf{u}_i \tag{3.4.14}$$

A sufficient condition for the second equation to hold is

$$f_i R_m^2(1/\omega_i) = u_{mi}^2 \tag{3.4.15}$$

for every $m = 0, 1, \ldots, n-1$ (let the elements of $\mathbf{u}_i$ also be numbered from 0 to $n-1$), and particularly

$$f_i R_0^2(1/\omega_i) = u_{0i}^2 \tag{3.4.16}$$

or

$$f_i = N_0 u_{0i}^2 \tag{3.4.17}$$

which is the desired expression for the quadrature weights.

The abscissae and weights have the following properties:

- All abscissae lie within the orthogonality interval $[a, b]$.
- Abscissae of successive orders mutually separate each other (interlace).
- All weights are nonnegative.
- The abscissae and weights of the quadrature of order n reproduce the first $2n$ moments

$$S(-k) = \sum_{i=1}^{n} f_i (1/\omega_i)^k \qquad k = 0, 1, \ldots, 2n-1 \tag{3.4.18}$$

In fact, they provide an exact integration if the integrated function is any polynomial in $1/\omega$ of degree lower than $2n$. Hence, if $2r$ moments are available ($k = 0, 1, \ldots, 2r-1$) quadrature parameters can be generated for quadratures of all orders n, with $n \leq r$.

3.5. Approximate distribution and density functions

Utilising the Gaussian quadrature abscissae ω_i and weights f_i an approximate distribution function $F^{(n)}(\omega)$ of order n may be constructed

$$\begin{aligned} F^{(n)}(\omega) &= 0 && \omega < \omega_1 \\ F^{(n)}(\omega) &= \sum_{j=1}^{i} f_j && \omega_i < \omega < \omega_{i+1} \\ F^{(n)}(\omega) &= \sum_{j=1}^{n} f_j = S(0) && \omega_n < \omega \end{aligned} \tag{3.5.1}$$

This is a nondecreasing staircase-like function and it is often referred to as the Stieltjes distribution. The points of increase (the "steps") lie at the quadrature abscissae ω_i. At these points the Stieltjes distribution satisfies the Chebyshev inequalities [1] (section 3.2.) which means that the plateaus to the right and to the left of the step provide the lower and upper bound to the value of the distribution function of infinite order $F(\omega_i)$.

By the above formulae the approximate distribution is defined only between the abscissae ω_i, but not at the abscissae themselves, where there are discontinuities. It is customary to take the value of $F^{(n)}(\omega_i)$ as the average between the the right-hand-side and the left-hand-side level at ω_i [33]

$$F^{(n)}(\omega_i) = \frac{1}{2}\left[F^{(n)}(\omega_i - 0) + F^{(n)}(\omega_i + 0)\right] \tag{3.5.2}$$

In the case of photoionisation the Stieltjes distribution $F^{(n)}(\omega)$ is equivalent to an approximate cumulative oscillator strength for the continuum range. Its differentiation leads to an approximate nth-order density function ("Stieltjes density") $f^{(n)}(\omega)$. Various recipes for performing this differentiation have been suggested. Hazi and Rescigno have fitted the discrete Stieltjes distribution by a continuous function which was then analytically differentiated [17, 39]. Broad and Reinhardt made use of the so-called Heller derivatives [30, 40]. The most popular approach is to differentiate the Stieltjes distribution numerically [9, 32, 33]

$$\begin{aligned} f^{(n)}(\omega) &= \frac{1}{2}\frac{f_1}{\omega_1} && \omega < \omega_1 \\ f^{(n)}(\omega) &= \frac{1}{2}\frac{f_{i+1} + f_i}{\omega_{i+1} - \omega_i} && \omega_i \leq \omega < \omega_{i+1} \\ f^{(n)}(\omega) &= 0 && \omega_n < \omega \end{aligned} \tag{3.5.3}$$

This yields the density function $f^{(n)}(\omega)$ as a nonnegative histogram function. It is then interpolated to give a continuous nth-order density. Cacelli et al. [41] least-square fitted a function to Stieltjes histograms of several orders at the same time. Müller-Plathe and Diercksen [42] used a cubic spline to interpolate the Stieltjes histogram of the highest possible order only. Monotonicity-constrained splines which avoid overshooting [43] have been found particularly useful.

It is also possible to obtain continuous distributions and densities ("Chebyshev" distributions and densities) directly without any interpolation or fitting if quasiorthogonal polynomials are used in place of the orthogonal polynomials [33]. This scheme, however, is numerically ill-conditioned except for very low order approximations. The reasons for this behaviour have been analysed by Müller-Plathe and Diercksen [31, 42] who concluded that the use of this method is to be discouraged.

At this stage the reduced moment problem may be regarded as solved. Starting from the raw transition energies $\bar{\omega}_i$ and oscillator strengths $\bar{f}_i$ approximations for the molecular continuum oscillator strength based on the generalised Gaussian quadrature abscissae ω_i and weights f_i have been determined which themselves are by their very nature kinds of transition energies and oscillator strengths. One may ask what has been gained by going through the whole procedure. The answer is that the cross section has been effectively smoothed. There are many more raw transitions (typically several hundred) than there are abscissae and weights (typically less than twenty). Because of the inaccuracy of the raw transitions and their basis-set dependence they can only describe correctly the global shape of the oscillator strength distribution but not its details. Using the raw transitions directly in the construction of the oscillator strength distribution without going through the moment-theory step would correspond to using all possible moments. This would yield a cross section which contains all the small errors in the transition parameters. They would show up as noise in the cross section. By using a much smaller number of "transitions" which are adapted to reproduce the first few moments only (which are reliable) we forfeit the information about small-scale features of the cross section (which is contained in the higher moments). This way of smoothing the cross section has the advantage over others like least-squares, splines or interpolations that the cross sections generated are by definition consistent with the sum rules which are observable quantities. The Stieltjes imaging process also places the abscissae where they are mostly needed, i.e. the density of abscissae is higher where the cross section changes most rapidly and lower where the cross section shows a smooth behaviour.

3.6. Related techniques

The "Stieltjes imaging" method for solving the moment problem which has been described in this chapter so far can be called a "numerical" method in the sense that it renders the distribution and density functions at discrete abscissae only. The discrete points can, of course, be interpolated to yield continuous representations. There also exist techniques in the literature [44] by means of which parameters of a continuous analytical function are adjusted in such a way as to reproduce the first few moments. Consider for example the Gaussian (normal) density

$$G(x) = N\frac{1}{\sigma\sqrt{2\pi}}\exp\big(-\frac{1}{2}(x-\bar{x})^2/\sigma^2\big) \tag{3.6.1}$$

The shape of this Gauss curve is completely determined by its first three moments: the zeroth moment gives the normalisation N, the first moment (the arithmetic mean $\bar{x}$)

gives the position of the maximum and the second moment (the variance σ^2) determines the dispersion (width) of the curve. From higher moments no more information can be extracted as there are no more adjustable parameters in the Gaussian density.

A more general technique is the expansion of the unknown density function $f(\omega)$ into a set of orthonormal polynomials $\{q_n(1/\omega)\}$ associated with a different, but known, density function $g(\omega)$ [44]

$$\int_a^b q_m(1/\omega)q_n(1/\omega)g(\omega)d\omega = \delta_{mn} \tag{3.6.2}$$

The integration limits have to be the same for $g(\omega)$ and $f(\omega)$. The expansion of $f(\omega)$ then takes the form

$$f(\omega) = g(\omega)\big(c_0q_0(1/\omega) + c_1q_1(1/\omega) + \cdots\big) \tag{3.6.3}$$

In order to determine the expansion coefficient c_n we multiply both sides of eqn. (3.6.3) with the orthogonal polynomial of degree n and integrate

$$\int_a^b q_n(1/\omega)f(\omega)d\omega = c_0\int_a^b q_0(1/\omega)q_n(1/\omega)g(\omega)d\omega +$$

$$c_1\int_a^b q_1(1/\omega)q_n(1/\omega)g(\omega)d\omega + \cdots + c_n\int_a^b [q_n(1/\omega)]^2g(\omega)d\omega + \cdots \tag{3.6.4}$$

Invoking orthonormality (eqn. 3.6.2) yields

$$c_n = \int_a^b q_n(1/\omega)f(\omega)d\omega \tag{3.6.5}$$

This integral can be rewritten in terms of the moments

$$c_n = q_n^0\int_a^b f(\omega)d\omega + q_n^1\int_a^b (1/\omega)f(\omega)d\omega + \cdots + q_n^n\int_a^b (1/\omega)^n f(\omega)d\omega \tag{3.6.6}$$

$$c_n = q_n^0S(0) + q_n^1S(-1) + \cdots + q_n^nS(-n) \tag{3.6.7}$$

where q_n^m again denotes the coefficient of the mth power in the orthonormal polynomial of degree n

In the so-called Gram-Charlier expansion the density function $g(\omega)$ is chosen to be a Gaussian and the corresponding polynomials are the Hermite polynomials. Expansions for other standard density functions can also be found [44]. The expansion into an analytical function generally converges very quickly if $g(\omega)$ is already very similar to $f(\omega)$ and the polynomial terms only have to account for a small perturbation. If $g(\omega)$ and $f(\omega)$ are very different then the expansion may become very long. In the case of photoionisation yet another problem arises. One would like to choose a $g(\omega)$ which has the same boundary conditions as $f(\omega)$. However, only the behaviour of $f(\omega)$ for large ω is known, where it approaches zero asymptotically. The behaviour of $f(\omega)$ at the ionisation threshold ω_T is not known a priori and may vary from case to case. The continuum oscillator strength may start with a finite value and decay slowly, or pass through a maximum. The above-threshold cross section may even be zero and then pass through a maximum. This may be the reason why analytical expansions have not been explored for photoionisation cross section calculations and the numerical Stieltjes method, which makes no assumptions at all about the form of $f(\omega)$, has been used exclusively.

4. Polarisation propagator methods

The moment-theoretical approach to the calculation of photoionisation cross sections relies on the availability of accurate moments of the continuum oscillator strength as has been described in the previous section. Even if the recurrence proceeds directly from the discrete transition energies and oscillator strengths and no intermediate moments are calculated the moments are still used implicitly and the transition parameters must reproduce them accurately. In this section we deal with quantum-chemical techniques which are capable of producing the required accurate moments within the framework of ordinary L^2 basis sets.

The method of choice is the polarisation propagator (PP) technique. Some of the approximations to the PP can be shown to satisfy certain sum rules of the total oscillator strength (discrete excitations plus continuum absorption, eqn. 2.8) in the limit of a complete basis set. Their theory is covered in detail by a number of review articles [45-48]. Here, only the aspects of the PP theory relevant to moment-theoretical calculations are briefly summarised.

The polarisation propagator method is derived within linear-response theory which deals with the response of a system to an external perturbation. In the case of the polarisation propagator the perturbation may be a time-dependent electric field in which case the propagator yields information about electric dipole transitions. Linear-response theory avoids the calculation of ground state and excited states wavefunctions, but obtains the transition energies and oscillator strengths directly from the poles and residues of the PP in contrast to the more familiar static approaches.

The lowest order PP approximation is equivalent to the configuration-interaction (CI) method that includes all single excitations (SCI) from a self-consistent field (SCF) ground state determinant. This method is also known as the multi-channel static-exchange (SX) approximation or Tamm-Dancoff approximation (TDA). Oscillator strengths obtained within the SX approximation in the dipole-length (DL) and dipole-velocity (DV) formulations are generally different and they do not satisfy any of the sum rules of section 2., even for a complete basis set. The violation of the DL-DV equivalence and the sum rules is a general feature of all incomplete CI methods. Extension of the SX method within the framework of CI methods is difficult, firstly because the sum rules converge slowly with the number of multiply-excited configurations included in the calculation, secondly because the large number of configurations which result quickly lead to Hamiltonian matrices that cannot be fully diagonalised. In moment theory, unlike in bound state calculations, all eigenstates are needed.

The first-order PP approximation, also known as random-phase approximation (RPA) or time-dependent Hartree-Fock method (TDHF) differs from the SX approximation in that the SCF determinant is no longer used as the electronic ground state but only as a reference state for the generation of the RPA matrix. The manifold of excitation operators now includes all single excitations and deexcitations. This leads to elements in the RPA matrix that correspond to double substitutions in addition to the terms arising from singly excited configurations that are already present in the SX matrix. The different formulations (DL, DV, DLDV) for the oscillator strengths are completely equivalent in RPA, and the TRK sum rule $s(0)$ is obeyed exactly, provided the basis set is large enough and all possible single particle-hole excitations are included. Note that the basis set need not include continuum functions for the TRK sum rule to be satisfied. A discrete L^2 basis, if sufficiently large, recovers all of the continuum oscillator strength. The static electric dipole polarisability and other second-order response properties are equal in RPA and in the coupled Hartree-Fock approximation within the same basis set. As the RPA oscillator strengths are identical in different dipole gauges, all oscillator strength moments become identical as well.

It has been noted previously that often more reliable results can be obtained in the

Table 1. Bibliography of moment-theoretical calculations of photoionisation cross sections which use transition energies and oscillator strengths from polarisation propagator calculations of at least first order in the electron correlation.

System	Method, Basis Set	Reference
Li^-, Na^-	RPA, effective core potential	[56]
H_2O	RPA, [5s 4p 1d; 5s 1p] CGTO	[57]
N_2	RPA, [N: 6s 5p 1d; cm: 7p 4d] CGTO	[58]
HF	RPA, [6s 10p 6d; 4s 4p] CGTO	[59]
HCl	RPA, large number of one-centre STO's	[60]
N_2, O_2	MCRPA, [7s 6p 4d] CGTO	[61]
H_2S	RPA, large number of one-centre STO's	[62]
Li^-	CCSDT1aPPA, (16s 10p) GTO	[55]
H_2O	SOPPA, (16s 9p 4d; 10s 3p) GTO	[31, 42]
Li^-	MCRPA, (10s 16p 9d) STO	[63]

SX approximation if the mixed length/velocity (DLDV) gauge rather than either the DL or the DV gauge is used [49, 50]. This feature is due to the cancellation of errors at the SX level in this gauge. It can be shown [51] that the oscillator strengths and hence the TRK sum rule calculated in the mixed gauge must be equal in SX and RPA, and in fact also in higher-order perturbative PP approximations. This equivalence holds only for $s(0)$, the transition energies obtained in different orders are generally different, so that other moments than $s(0)$ also differ.

Of the linear response methods that go beyond the RPA level only the multiconfigurational extension of RPA (MCRPA or MCTDHF) provides a strict dipole-length/dipole-velocity equivalence and only MCRPA obeys the TRK sum rule exactly in all gauges. The reason for this is that only RPA, MCRPA and full-CI PP satisfy the time-dependent analog of the Hellmann-Feynman theorem [23]. For the perturbative PP Jørgensen and Oddershede [52] have shown that the equivalence restrictions and the sum rules are fulfilled through to the order in perturbation theory to which the PP is evaluated consistently. This is in marked contrast to the SX approximation which may produce large differences between oscillator strengths calculated in different gauges and errors in the TRK sum rule are often of the order of 25 % or more.

The second-order PP approximation (SOPPA) uses as reference state a manybody perturbation theory (MBPT) expansion which contains double substitutions in first order and single substitutions in second order. All elements of the PP matrix and all contributions to the transition moments are evaluated consistently through second order in perturbation theory so that transition energies and oscillator strengths are consistent through this order. To achieve a consistent second-order treatment terms involving singly, doubly and triply substituted configurations have to be included implicitly into the SOPPA method [53].

Higher-order MBPT-based polarisation propagator methods have not been implemented so far. However, the SOPPA method has been extended to coupled-cluster (CC) reference states [48, 54]. In connection with moment theory only the CCSDT1a variant (CC including single, double and linearised triple excitations) of the coupled-cluster PP approximation has been used [55].

Although the shortcomings of the SX method have been known since long the vast majority of all moment-theoretical calculations have employed transition energies and oscillator strengths obtained in this approximation. The fact that this approximation does not reproduce the TRK sum rule has often led to photoionisation cross sections which are globally over- or underestimated. Transition parameters from RPA calculations are certainly the minimum requirement for use in photoionisation cross section calculations. To our knowledge there have been only a few photoionisation cross section calculations reported in the literature which employ RPA or higher PP approximations. The references to these calculations are compiled in table 1.

There have also been some calculations of atomic and molecular oscillator strength distribution moments and mean excitation energies by various polarisation propagator approximations. Hence, the performance of the different PP techniques and their requirements on the basis set are fairly well established (see references. [31, 42, 65-67] and references therein).

5. Selection of basis sets

In the foregoing chapters basis sets have been mentioned a few times. A basis set for a moment-theoretical cross section calculation has to meet two requirements. Firstly, the basis set, which by definition consists of square-integrable L^2 functions such as Gaussian-type orbitals (GTO's) or Slater-type orbitals (STO's) has to be able to mimic the behaviour of the continuum functions in the molecular region.

Hence, there must be a considerable flexibility in the basis to accomplish just that. Secondly, some features of certain polarisation propagator approximations hold only in the limit of a complete basis set, such as the reproduction of the TRK sum rule and the DL/DV equivalence restriction. The TRK sum rule provides the normalisation of the cross section and is therefore vital to the calculation. The DL/DV equivalence is an important probe of the practical completeness of the basis. In other words, it is more reassuring if oscillator strengths, moments and photoionisation cross sections obtained in the two gauges agree than if they don't.

The photoionisation cross section is calculated from the moments $S(0)$, $S(-1)$, ..., $S(-(2r-1)$. Therefore their dependence on different parts of the oscillator strength distribution is discussed first. The TRK sum rule depends equally on the whole absorption spectrum. In contrast, the polarisability is completely determined by the discrete excitations and the low-energy continuum. Negative sum rules depend mainly on the low-energy part of the spectrum because the energies enter with inverse powers. For moments or mean excitation energies that contain positive powers or functions of ω the opposite is true. They depend mainly on the high-energy continuum part of the spectrum. Thus different sum rules are sensitive to different parts of the spectrum as is illustrated by the example of the H_2O molecule [64]: The discrete part of the spectrum below the first ionisation threshold of 12.61 eV contributes only 3 % to the TRK sum rule, the rest is contributed by the continuum. On the other hand, the discrete part alone contributes 25 % to the polarisability. Consequently, the calculation of different sum rules imposes different requirements on the basis sets. A balanced description of the whole spectrum, including the high-energy continuum, is important in the calculation of the TRK moment. The basis must describe the discrete spectrum and the low-energy continuum very well when higher negative moments are calculated. If the moments are to be used in a photoionisation cross section calculation the basis has to fulfill both requirements.

There are essentially two approaches to devising basis sets for photoionisation calculations (see also table 1.). The Pisa group [60, 62, 63] used one-centre STO's. They first performed an SCF target state calculation in a fairly large basis and then projected the solution onto a smaller basis set in order to reduce the number of orbitals. This basis is then augmented by STO's with small exponents for the valence excited states, hydrogen-like orbitals with appropriate quantum defects for the Rydberg series and so-called oscillating STO's for the continuum. The latter contain a $\cos(kr)$ term in the radial part to account for the oscillations of the continuum orbitals. The total number of basis functions is usually well over one hundred. The one-centre STO expansion is particularly suitable for systems consisting of a central heavy atom decorated with hydrogens, in which case the heavy atom is taken as the origin of the STO's. A generalisation to other systems might be difficult.

The other approach uses a large number of GTO's. This has the advantage that the multicentre nature of both the target and the continuum wave functions can be easily allowed for. The bases are constructed in the following way [31, 42]. For description of the target one starts with a well polarised, highly contracted basis of about triple-ζ quality which may be augmented with diffuse functions. Then more diffuse functions of appropriate symmetry are added to describe the excited (pseudo)states. Bases of this type should saturate the static polarisability and the higher negative moments, as the excited states and the low continuum are covered. However, the high-energy continuum is still unaccounted for and therefore the TRK sum rule as well as $s(-1)$ and $I(0)$ are not satisfied. The high-energy continuum functions oscillate rapidly and therefore very compact basis functions have to be included. An inexpensive way of achieving this is to uncontract the initial target basis. For this reason a highly contracted basis set is chosen initially. Different centres have been tried for the diffuse continuum basis. For hydrides it appears to be reasonable to place most of the continuum functions on the heavy atom [31, 42, 59]. For homonuclear diatomics it seems to be more economic to put the continuum functions only on the centre of mass [58]. In either way, the Gaussian bases have to be large. The RPA results for the water molecule obtained in a small basis set are qualitative estimates only [57].

6. Some practical considerations

As discussed in section 3.5. the Stieltjes imaging procedure is a smoothing method because it uses only a few of all available moments. Hence, the resolution of the method is limited. Very sharp features such as certain resonances, autoionising states or vibrational couplings cannot be resolved. The smoothing simply integrates them into the background continuum. Therefore, in regions where there are many spikes in the cross section (say below 20 eV where usually excitations from some channels interact with the ionisation continuum of other channels) Stieltjes imaging provides only averaged information about the cross sections. It becomes considerably more accurate in the region above 20-25 eV and therefore complements methods for the calculation of the low-energy region with all its details resolved such as the R-matrix method [68]. The typical resolution of the Stieltjes method is several eV so that broad shape resonances can be detected.

The next point to be addressed is the coupling of channels. A channel is understood as the ionisation out of a particular orbital. If the total ionisation can be sensibly decomposed into contributions from different channels (consult, for instance, the molecule's photoelectron spectrum at different photon energies) it makes sense to calculate partial cross sections for the individual channels. This is done in the polarisation propagator calculation. One may restrict the occupied orbital space to the orbital presently under consideration [42]. This is called the separated-channel approximation. Alternatively, one can perform a calculation including all occupied orbitals and then project the transitions onto single particle-hole excitations. The extent of coupling between different channels in a given transition may also be judged by inspecting the contributions of the individual occupied-to-virtual excitations. However, configuration mixing may be overemphasised if the basis set is not large enough. Then the mixing accounts for deficiencies in the one-electron basis rather than being physically meaningful.

The total cross section can be obtained within the separated-channel approximation by adding up the cross sections for the individual channels. Apart from neglecting the physical effects due to channel coupling the total cross section thus generated exhibits artificial discontinuities at each ionisation potential. For the calculation of more realistic cross sections it is therefore necessary to give up the separated-channel approximation and treat all channels simultaneously. Only the core excitations may usually be separated to a good approximation from the valence excitations. Performing a fully coupled calculation is usually not a problem at the RPA level, but becomes a serious computational challenge at higher levels of PP theory. Since the cross sections usually show a very regular behaviour at high energies an RPA description of that part of the spectrum is sufficient in most cases. It is therefore often a good approximation to calculate the excitations which cover the energy range that determines the negative moments, say up to 100 eV, by the higher-order method (MCRPA, SOPPA, etc.) and to fill in the rest of the spectrum with transitions from an RPA calculation.

Another problem arises in the calculation of cross sections for inner-valence-shell and core orbitals. The hole left in such an orbital by the photoionisation process is a much larger perturbation to the other electrons than a hole in an outer-valence orbital.

The PP calculation is usually performed in the space of canonical SCF orbitals. Therefore the effect of orbital relaxation upon removal of an inner electron is not accounted for in the one-electron basis but has to be recovered by the inclusion of other electronic states. The limited treatment of electron correlation in all PP approximations leaves a noticeable error in the calculated transition energies for inner orbitals. The possibility of using a modified-virtual-orbital (MVO) space has been investigated [69]. It has no effect at the RPA level because RPA is invariant to rotations among either the occupied or the virtual orbitals. At higher levels (SOPPA) the use of MVO's has been found to make results worse rather than improve them [69]. For PP approximations that are not too advanced in the treatment of electron correlation the error may be estimated to be equal to the difference between the experimental ionisation potential and its estimate from Koopmans' theorem [31, 42]. A better agreement between experimental and calculated cross sections can be obtained by shifting the transitions by this empirical correction assuming that the error introduced is approximately equal for all transitions out of the inner orbital. These considerations apply only to the calculation of partial cross sections for inner-valence-shell and core channels. As these cross sections are much smaller than the outer-valence cross sections the effect of the corrections on the total photoionisation cross section is marginal.

Finally, the question remains which order of approximation should be used in the Stieltjes imaging. The higher the order the better the resolution is, but the more inaccurate the moments become. Looking at the cross sections of increasing orders one sees initially a rather structureless curve in which gradually the features take shape before at some order the graph shows an unphysical behaviour. This is certainly beyond the point to stop. However, this inspection is not a satisfactory way of determining the order. A more rigorous criterion can be established by considering the behaviour of the orthogonal polynomial recurrence coefficients in the limit of an infinite order of approximation [7, 32]

$$a_\infty = \frac{1}{2\omega_T} \tag{6.1}$$

$$b_\infty = \frac{1}{(4\omega_T)^2} \tag{6.2}$$

Examining the recurrence coefficients one typically finds that after some initial oscillations they begin to converge to their infinite-order limits, until at a certain order they suddenly diverge. This indicates the point where the available moments become too inaccurate to sustain a reliable cross section.

Acknowledgments

FMP would like to thank the organisers of the "II Escola Brasileira de Estrutura Eletrônica", especially Sylvio Canuto, for the great hospitality at Recife and The British Council for financial support. GHFD would like to thank the Fonds der Chemischen Industrie im Verband der Chemischen Industrie e.V. for partial support. Many helpful discussions with Jens Oddershede and Jacek Karwowski are gratefully acknowledged.

References

[1] J. A. Shohat and J. D. Tamarkin, *The problem of moments (Math. surveys I)* (American Mathematical Society, Providence, 1943).

[2] N. I. Akhiezer, *The classical moment problem (and some related questions in analysis)* (Oliver and Boyd, Edinburgh and London, 1965).

[3] G. Szegö, *Orthogonal polynomials*, 4th edition (American Mathematical Society, Providence, 1975).

[4] G. Freud, *Orthogonal polynomials* (Pergamon, Oxford, 1971).

[5] H. S. Wall, *Analytic theory of continued fractions* (van Nostrand, New York, 1948; reprinted by Chelsea, New York, 1973).

[6] P. J. Davis and P. Rabinowitz, *Methods of numerical integration*, 2nd edition (Academic Press, Orlando, 1984).

[7] R. G. Gordon, Adv. Chem. Phys. *15*, 79 (1968).

[8] R. G. Gordon, J. Math. Phys. *9*, 655 (1968).

[9] J. Deltour, Physica *39*, 413 (1968); ibid. 424.

[10] J. C. Wheeler and R. G. Gordon, J. Chem. Phys. *51*, 5566 (1969).

[11] C. Blumstein and J. C. Wheeler, Phys. Rev. B *8*, 1764 (1973).

[12] M. Roseau, *Vibration in mechanical systems – analytical methods and applications* (Springer, Berlin, 1987).

[13] P. W. Langhoff and M. Karplus, in G. A. Baker Jr. and J. R. Gammel (eds.) *The Padé approximant in theoretical physics* (Academic Press, New York, 1970).

[14] P. W. Langhoff and M. Karplus, J. Chem. Phys. *52*, 1435 (1970).

[15] P. W. Langhoff, J. Chem. Phys. *57*, 2604 (1972).

[16] P.W. Langhoff and S. L. Seidman, Chem. Phys. Lett. *27*, 195 (1974).

[17] A. U. Hazi, J. Phys. B *11*, L259 (1978).

[18] H. Ågren and V. Carravetta, J. Chem. Phys. *87* (1987).

[19] C. M. M. Nex, J. Phys. A *11*, 653 (1978).

[20] M. R. Hermann and P. W. Langhoff, J. Math. Phys *24*, 541 (1983); Phys. Rev. A *28*, 1957 (1983); Int. J. Quantum Chem. *23*, 135 (1983).

[21] P. W. Langhoff, in G. H. F. Diercksen and S. Wilson (eds.), *Methods in computational molecular physics* (D. Reidel, Dordrecht, 1983).

[22] K. Kaufmann, W. Baumeister and M. Jungen, J. Phys. B *20*, 4299 (1987).

[23] A. J. Sadlej, Int. J. Quantum Chem. *35*, 409 (1989).

[24] H. A. Bethe and E. E. Salpeter, *Quantum mechanics of one- and two-electron atoms* (Springer, Berlin, 1957; reprinted by Plenum, New York, 1977).

[25] J. O. Hirschfelder, W. Byers Brown and S. O. Epstein, Adv. Quant. Chem. *1*, 255 (1964).

[26] U. Fano and J. W. Cooper, Rev. Mod. Phys. *40*, 441 (1968).

[27] H. A. Bethe, Annalen der Physik (Leipzig) *5*, 325 (1930).

[28] M. Inokuti, Rev. Mod. Phys. *43*, 297 (1971).

[29] H. A. Bethe and R. Jackiw, *Intermediate quantum mechanics*, 3rd edition (Benjamin Cummings, Menlo Park, 1986).

[30] W. P. Reinhardt, Comput. Phys. Commun. *17*, 1 (1979).

[31] F. Müller-Plathe, PhD thesis (Technische Universität, München, 1988); Report MPA 368 (Max-Planck-Institut für Astrophysik, Garching, 1988).

[32] P. W. Langhoff, C. T. Corcoran, J. S. Sims, F. Weinhold and R. M. Glover, Phys. Rev. A *14*, 1042 (1976).

[33] C. T. Corcoran and P. W. Langhoff, J. Math. Phys. *18*, 651 (1977).

[34] W. H. Press, B. P. Flannery, S. A. Teukolsky, and W. T. Vetterling, *Numerical recipes* (Cambridge University Press, Cambridge, 1986).

[35] M. Ryšavý, Comput. Phys. Commun. *32*, 191 (1984).

[36] T. Takemasa, Comput. Phys. Commun. *48*, 265 (1988); ibid. *52*, 133 (1988).

[37] F. Müller-Plathe, Information quarterly for computer simulation of condensed phases (CCP5 newsletter) No. 30 (Daresbury Laboratory, Warrington, April 1989).

[38] e.g. TQL2 or IMTQL2 from the EISPACK package, see B. T. Smith, J. M. Boyle, J. J. Dongarra, B. S. Garbow, Y. Ikebe, V. C. Klema and C. B. Moler, *Matrix eigensystems routines – EISPACK guide*, Lecture Notes in Computer Science Vol. 6 (Springer, Berlin, 1976); B. S. Garbow, J. M. Boyle, J. J. Dongarra and C. B. Moler, *Matrix eigensystems routines – EISPACK guide extensions*, Lecture notes in computer science Vol. 51 (Springer, Berlin, 1976).

[39] A. U. Hazi and T. N. Rescigno, Phys. Rev. A *16*, 2376 (1977).

[40] J. T. Broad and W. P. Reinhardt, Chem. Phys. Lett. *37*, 212 (1976).

[41] I. Cacelli, V. Carravetta and R. Moccia, Chem. Phys. *90*, 313 (1984).

[42] F. Müller-Plathe and G. H. F. Diercksen, Phys. Rev. A *40*, 696 (1989).

[43] J. M. Hyman, SIAM J. Sci. Stat. Comput. *4*, 645 (1983).

[44] H. Cramér, *Mathematical methods of statistics* (Princeton University Press, Princeton, 1966).

[45] J. Oddershede, Adv. Chem. Phys. *69*, 201 (1987).

[46] J. Oddershede, Adv. Quant. Chem. *11*, 275 (1978).

[47] J. Oddershede, P. Jørgensen and D. L. Yeager, Comput. Phys. Reports *2*, 33 (1984).

[48] J. Geertsen, Adv. Quant. Chem. (to be published).

[49] I. Cacelli, V. Carravetta and R. Moccia, Mol. Phys. *59*, 385 (1986).

[50] I. Cacelli, V. Carravetta and R. Moccia, Chem. Phys. *120*, 51 (1988).

[51] P. Jørgensen, J. Oddershede and N. H. F. Beebe, J. Chem. Phys. *68*, 2527 (1978).

[52] P. Jørgensen and J. Oddershede, J. Chem. Phys. *78*, 1898 (1983).

[53] J. Oddershede and P. Jørgensen, J. Chem. Phys. *66*, 1541 (1977).

[54] J. Geertsen and J. Oddershede, J. Chem. Phys. *85*, 2112 (1986).

[55] S. Canuto, J. Geertsen, F. Müller-Plathe and G.E. Scuseria, J. Phys. B *21*, 3891 (1988).

[56] R.F. Stewart, C. Laughlin and G.A. Victor, Chem. Phys. Lett. *29*, 353 (1974).

[57] G.R.J. Williams and P.W. Langhoff, Chem. Phys. Lett. *60*, 201 (1979).

[58] G.R.J. Williams and P.W. Langhoff, Chem. Phys. Lett. *78*, 21 (1981).

[59] A.P. Hitchcock, G.R.J. Williams, C.E. Brion and P.W. Langhoff, Chem. Phys. *88*, 65 (1984).

[60] I. Cacelli, V. Carravetta and R. Moccia, Mol. Phys. *59*, 385 (1986).

[61] P. Swanstrøm, J.T. Golab, D. L. Yeager and J.A. Nichols, Chem. Phys. *110*, 339 (1986).

[62] I. Cacelli, V. Carravetta and R. Moccia, Chem. Phys. *120*, 51 (1988).

[63] V. Carravetta, H. Ågren, H. J. Aa. Jensen, P. Jørgensen and J. Olsen, J. Phys. B *22*, 2133 (1989).

[64] J. Berkowitz, *Photoabsorption, photoionization, and photoelectron spectroscopy* (Academic Press, New York, 1979).

[65] J. Geertsen, J. Oddershede and J. R. Sabin, Phys. Rev. A *34*, 1104 (1986).

[66] S. Canuto, W. Duch, J. Geertsen, F. Müller-Plathe, J. Oddershede and G. E. Scuseria, Chem. Phys. Lett. *147*, 435 (1988).

[67] J. Oddershede and J. R. Sabin, Phys. Rev. A *39*, 5565 (1989).

[68] J. Tennyson, C. Noble and P. G. Burke, Int. J. Quantum Chem. *29*, 1033 (1986).

[69] J. Oddershede and J. R. Sabin, J. Chem. Phys. *79*, 2295 (1983).

EXTENSION OF DENSITY FUNCTIONAL THEORY TO EXCITED STATES

L. N. Oliveira

Instituto de Física e Química de São Carlos
Universidade de São Paulo
13560 São Carlos, SP—BRAZIL and
Physics Department—University of California
Santa Barbara, CA 93106—USA

ABSTRACT

Recent developments aimed at extending Density Functional Theory to excited states are described. For completeness, the review starts with the fundamentals of ground-state Density Functional Theory, but most of the paper is dedicated to the ensemble approach, first suggested by Theophilou ten years ago, in which the excitation energies are calculated from the average energies of ensembles containing a finite number of excited states. The Rayleigh- Ritz principle for excited states formulated by Gross *et al.*, which gives the ensemble approach a broader foundation, is discussed. The density-functional formalism constructed on the basis of that principle is described. Kohn's quasi-local density approximation, which turns that formalism into a practical calculational procedure is also presented. Finally, a test application of the formalism that calculates the ten lowest excitation energies for the He atom is discussed and compared to the Slater transition state approach.

1. INTRODUCTION

Density functional theorists are currently commemorating the 25th anniversary of the Hohenberg-Kohn paper[1]. Since that cornerstone, a succession of developments have turned the density functional approach into a powerful technique to treat the electron-electron interaction in atoms, molecules, and solids. In particular, in the last decade substantial progress has been made towards adapting the theory—originally devised as a ground-state formalism—to the calculation of excitation energies of small systems (atoms and molecules)[2,3,4]. This chapter reviews the central aspects of this more recent development.

In an effort to make the review self-contained, I start out by reproducing the fundamental results at the basis of ground-state Density Functional Theory. This lead in is meant to be neither deep nor comprehensive. For discussions of the implications of the Hohenberg-Kohn theorem and of its ramifications the reader is referred to one of many excellent reviews[5]. A particularly instructive survey aimed at the uninitiated reader has been provided by Pedroza in the Proceedings of the first Brazilian School on Electronic Structure[6]. Here, as the title dictates, the extension of the formalism to excited states will be emphasized.

Several alternative generalizations of the groud-state formalism to excited states have been proposed. Examples are the Green's function approach, started by Sham and Kohn[7], and the application of the ground-state formalism to the lowest state of a symmetry class[8,9].

Another method, contemporary to the work of Sham and Kohn[7], is due to Mermin[10], who extended the Rayleigh-Ritz principle to the grand-canonical ensemble and the Hohenberg-Kohn theorem to finite temperatures. Given that currently available[11] expressions for the thermodynamical properties of the homogeneous electron gas generate a satisfactory local-density approximation (LDA) for the thermal exchange-correlation energy functional, Mermin's work has provided a practical extension of the Hohenberg-Kohn-Sham procedure. Unfortunately, to calculate excitation energies this method is unwieldy: one would have first to calculate the partition function Z at all temperatures and then to invert the expression $Z = \sum_j \exp(-\beta E_j)$ for the energies E_j.

The extension discussed in this chapter follows a more recent approach to the calculation of excited-state energies, initiated by Theophilou[2], who considered the ensemble of the M lowest states, equally weighted. For this *equiensemble*, an extension of the Rayleigh-Ritz principle[12] had long been available, enabling him to generalize the Hohenberg-Kohn-Sham theory, so that the ensemble averaged energies and densities can (at least formally) be determined. Once these arithmetic averages are known for $M = 1, \ldots, I$, the lowest I individual energies and the corresponding densities are easily found by solving a linear system of M equations.

This work has started a new branch in Density Functional Theory. A recent development along this line has extended the Rayleigh-Ritz principle to ensembles of *unequally* weighted states and the Hohenberg-Kohn-Sham scheme to ensembles of fractionally occupied states[4,13]. This extended formalism contains Theophilou's equiensemble approach to the calculation of excited-state energies as a special case. In addition, it provides an alternative procedure (the *fractional-occupation* approach), which relates excitation energies to Kohn-Sham (single-particle) eigenvalues[14]; the latter method is similar to Slater's transition state method[15], a procedure involving densities computed from fractionally occupied orbitals and excitation energies calculated from single-particle eigenvalues of the Hartree-Fock-Slater equations.

This review is aimed at deriving the fundamentals of both the *equiensemble* and the *fractional occupation* approaches and describing their application to a simple problem, that of the He atom. The discussion is organized as follows. The basics of Density Functional Theory, the Hohenberg-Kohn theorem and the derivation of the Kohn-Sham equations, are presented in Section 2.. The local-density approximation (LDA) for the exchange-correlation energy functional, a time-honored means of converting the formally exact ground-state Kohn-Sham equations into powerful, practical calculational tools, is described in Section 3.. The remaining sections are dedicated to excited states. Section 4. extends the Rayleigh-Ritz principle to ensembles comprising the ground state and excited states. On the basis of this generalized principle, an ensemble density-functional formalism is discussed in Section 5., which ultimately leads to (formally exact) explicit expressions for excitation energies. To become practical, those expressions require an approximation for the ensemble exchange-correlation energy functional. A *quasi-local* form, recently proposed by Kohn[3], is presented in Section 6.. Section 7. applies the results of Sections 5. and 6. to the He atom, whose ten lowest excited-state energies are compared with experimental results and with the results of the Slater transition-state method. Finally, Section 8. collects general conclusions.

2. GROUND-STATE DENSITY FUNCTIONAL THEORY

Density functional theory is concerned with the electron-electron interaction defined by the Hamiltonian

$$\begin{aligned} H &= \int \psi^\dagger(r)\frac{-\nabla^2}{2}\psi(r)\,dr + \int \psi^\dagger(r)v(r)\psi(r)\,dr \\ &+\frac{1}{2}\int\int \psi^\dagger(r)\psi^\dagger(r')\frac{1}{|r-r'|}\psi(r')\psi(r)\,dr\,dr', \end{aligned} \tag{1}$$

in atomic units.

The diagonalization of this Hamiltonian, made difficult by the Coulomb interaction in the second term on the right-hand side, is further complicated by the external potential $v(r)$, which breaks the translational invariance of H. For small number N of particles, the Rayleigh-Ritz principle provides accurate approximations for the ground state energy. For large N, however, the many-particle wavefunction creates enormous difficulties. To avoid them, density functional theory establishes a variational principle involving the ground-state density

$$\rho(r) = \langle\Omega|\psi^\dagger(r)\psi(r)|\Omega\rangle \tag{2}$$

instead of the wavefunction. This principle is founded on the Hohenberg-Kohntheorem, which states that $\rho(r)$ determines the potential $v(r)$ in Eq. (1) up to within a constant. The following proof neglects the v-representability problem, for discussions of which the reader is referred to the Hohenberg-Kohn paper[1] and to those by Levy[16] and Lieb[17].

The proof proceeds by *reductio ad absurdum*. Assume, therefore, that two potentials $v(r)$ and $v'(r)$, corresponding to Hamiltonians H and H' according to Eq. (1), yield the same ground-state density $\rho(r)$. As long as $v(r)$ and $v'(r)$ differ by more than a constant, these Hamiltonians must have different ground states $|1\rangle$ and $|1'\rangle$, respectively. The Rayleigh-Ritz principle then implies the strict inequality

$$\langle 1|H|1\rangle < \langle 1'|H|1'\rangle, \tag{3}$$

the state $|1'\rangle$ considered a trial state for the Hamiltonian H.

Letting V (V') denote the operator $\int \psi^\dagger(r)v(r)\psi(r)\,dr$ [$\int \psi^\dagger(r)v(r)\psi(r)\,dr$], we therefore find

$$\langle 1|H|1\rangle < \langle 1'|H'|1'\rangle + \langle 1|V|1\rangle - \langle 1'|V'|1'\rangle, \tag{4}$$

or

$$\langle 1|H|1\rangle < \langle 1'|H'|1'\rangle + \int \rho(r)v(r)\,dr - \int \rho(r)v'(r)\,dr. \tag{5}$$

If $|1\rangle$ is now considered a trial state for the Hamiltonian H', the reasoning that led from Eq. (3) to Eq. (5) yields

$$\langle 1'|H'|1'\rangle < \langle 1|H|1\rangle + \int \rho(r)v'(r)\,dr - \int \rho(r)v(r)\,dr. \tag{6}$$

By adding this inequality to (5) we are then led to the inconsistency

$$0 < 0,$$

showing that the density $\rho(r)$ determines the potential $v(r)$.

It follows immediately from this result that given the Hamiltonian H (1) and given a density $\rho'(r)$, the latter determines a potential $v'(r)$, hence the ground-state $|1'\rangle$ of the Hamiltonian $H' = H + V' - V$, hence the expectation value

$$E_v[\rho'(r)] = \langle 1'|H|1'\rangle, \tag{7}$$

where the subscript v on the left-hand side reminds us that the expectation value of the Hamiltonian H, not of H', is being taken on the right-hand side. $E_v[\rho'(r)]$ is a functional of $\rho'(r)$ whose minimum—according to the Rayleigh-Ritz principle—must be $E_v[\rho(r)]$ [here $\rho(r)$ is the density associated with the potential $v(r)$], *i.e.*,

$$E_v[\rho(r)] < E_v[\rho'(r)] \text{ for } \rho(r) \neq \rho'(r). \tag{8}$$

This is the variational principle for the density upon which ground-state Density Functional Theory is founded. To take advantage of this principle, an approximation for $E_v[\rho'(r)]$ is necessary, a question discussed in Section 3.. For the moment, let us assume that $E_v[\rho']$ is known, to see how the Kohn-Sham construction[18] determines $\rho(r)$ from Eq. (8).

Parallel to the interacting system in which one is interested, Kohn and Sham considered a *non-interacting* electron gas with Hamiltonian

$$H_s = \int \psi^\dagger(r)\frac{-\nabla^2}{2}\psi(r)\,dr + \int \psi^\dagger(r)v_s(r)\psi(r)\,dr, \tag{9}$$

where the potential $v_s(r)$ is chosen to make the density of the non-interacting gas identical to that of the interacting gas with Hamiltonian H [Eq. (1)]. According to the Hohenberg-Kohn theorem, v_s is a unique functional of the density, $v_s(r) = v_s[\rho'](r)$. If $v_s[\rho']$ were known, the density could be easily found, by solving the Schrödinger Equation associated with H_s,

$$[-\frac{\nabla^2}{2} + v_s(r)]\phi_i(r) = \epsilon_i\phi_i(r) \qquad (i = 1, \ldots, N) \tag{10}$$

to determine the N lowest single-particle wavefunctions $\phi_i(r)$ and eigenvalues ϵ_i, and then computing the density as

$$\rho(r) = \sum_{i=1}^{N} |\phi_i(r)|^2. \tag{11}$$

$v_s[\rho']$ can be determined from $E_v[\rho']$ (assumed known). To see this, define with Kohn and Sham the exchange-correlation functional

$$E_{xc}[\rho'] = E_v[\rho'] - \int \rho'(r)v(r)\,dr - T_s[\rho'] - \frac{1}{2}\int\int \frac{\rho'(r)\rho'(r')}{|r-r'|}\,dr\,dr,' \tag{12}$$

where $T_s[\rho']$, a functional of $\rho'(r)$, is the kinetic energy associated with the ground state of the non-interacting gas, *i.e.*,

$$T_s[\rho'] = \sum_{i=1}^{N} \int \phi_i^*(r)\left(-\frac{\nabla^2}{2}\right)\phi_i(r)\,dr, \tag{13}$$

or according to Eqs. (10) and (11),

$$T_s[\rho'] = \sum_{i=1}^{N} \epsilon_i - \int v_s(r)\rho'(r)\,dr, \tag{14}$$

which takes advantage of the normalization of the single-particle wavefunctions $\phi_i(r)$.

Since $E_v[\rho']$ depends on v only explicitly, through the second term on the right-hand side of Eq. (1), the difference $E_v[\rho'] - \int v(r)\rho'(r)$ is universal, *i.e.*, a functional independent of $v(r)$. The definition (12) guarantees $E_{xc}[\rho']$ to be likewise universal. As shown in Section 3., this universality allows one to develop a simple approximation for $E_{xc}[\rho']$.

This considered, it is convenient to rewrite Eq. (12) as an expression that, given $E_{xc}[\rho']$, determines the energy functional $E_v[\rho']$:

$$E_v[\rho'] = T_s[\rho'] + \int \rho'(r)v(r)\,dr + \frac{1}{2}\int\int \frac{\rho'(r)\rho'(r')}{|r-r'|}\,dr\,dr,' + E_{xc}[\rho'], \tag{15}$$

or, account taken of Eq. (14),

$$\begin{aligned} E_v[\rho'] &= \sum_{i=1}^{N} \epsilon_i + \int \rho'(r)\{v(r) - v_s[\rho'](r)\}\,dr \\ &\quad + \frac{1}{2}\int\int \frac{\rho'(r)\rho'(r')}{|r-r'|}\,dr\,dr,' + E_{xc}[\rho']. \end{aligned} \tag{16}$$

It follows from the variational principle [Eq. (8)] that the functional derivative of the right-hand side of this equation with respect to ρ' vanishes at the correct density $\rho'(r) = \rho(r)$. When this functional derivative is calculated, with help of the result $\delta\epsilon_i/\delta\rho' = \int |\phi_i|^2 \delta v_s[\rho'](r)/\delta\rho'(r)\, dr$ [which follows from a standard result of first order perturbation theory applied to Eq. (10)], and then equated to zero, one finds that

$$v_s[\rho'](r) = v(r) + \int \frac{\rho(r)}{|r - r'|}\, dr + v_{xc}[\rho'](r) \tag{17}$$

where the exchange-correlation potential is the functional derivative

$$v_{xc}[\rho'](r) = \frac{\delta E_{xc}[\rho']}{\delta\rho'(r)}. \tag{18}$$

Eq. (17) resembles the potential in Hartree-Fock theory; in particular the first two terms on the right-hand side reproduce the Hatree terms of the latter. By contrast, the third term, a local potential, differs from the non-local exchange term characteristic of the Hartree-Fock method. Since it has long been known that non-locality is an essential ingredient of the latter approach, it is at first sight surprising to find a local potential in the formally exact Kohn-Sham equations. A longer analysis nevertheless shows that this local potential is a non-local functional of the density; this derives from Eq. (18), for although the *exact* form of E_{xc} is unknown, one expects it to be a strongly non-local functional of ρ' (the adjective ''exact'' has been emphasized because, as one might guess from its name, the LDA specifies a local form for E_{xc}).

Eqs. (17) and (18) complete the set of Kohn-Sham equations that Eqs. (10) and (11) started. To solve them one first considers an approximation for $v_s(r)$ [say $v_s(r) = v(r)$]. The Schrödinger Equation (10) is then solved and the density computed from Eq. (11). Eqs. (17) and (18) then yield an improved approximation for v_s, thus completing one iterative loop. The process is repeated until self-consistency is achieved within a desired accuracy. Eq. (15) then determines the ground state energy.

To implement this program numerically is straightforward. Nonetheless, since Eq. (17) [and Eq. (15)] takes $E_{xc}[\rho']$ as input, an approximate form for the exchange-correlation energy functional is indispensable. One is faced here with an essential difficulty in Density Functional Theory, for lacking one such approximation the Kohn-Sham equations would be reduced to formal identities. Fortunately, the LDA, discussed next, provides a practical solution to this problem.

3. THE LOCAL DENSITY APPROXIMATION

The local-density approximation, first proposed by Kohn and Sham [18], provides a simple form---one that can be easily computed on the basis of expressions for the ground-state energy of the *uniform* electron gas---for the exchange-correlation energy functional. The off-hand-unexpected success of this approximate form, the subject of several investigations[8,19], is to a large extent responsible for the popularity of the density functional approach in condensed matter theory. The reader is referred to those papers for discussions that justify the approximation. For the purposes of the present paper, a much briefer description will suffice.

Consider a system whose density varies slowly. It is then a good approximation to write the exchange-correlation energy functional in the local form

$$E_{xc}[\rho] \cong \int e_{xc}[\rho(r)]\,dr. \tag{19}$$

Here $e_{xc}[\rho(r)]$, the exchange-correlation energy per unit volume of a uniform electron gas, is computed as follows. For each position r, one considers a *uniform* interacting electron system with density ρ equal to the local density $\rho(r)$. Given an approximation for the energy of this uniform system, one can then calculate its exchange-correlation energy from Eq. (3). $e_{xc}[\rho(r)]$ is then computed as this exchange-correlation energy divided by the volume of the uniform system.

The local form of Eq. (19) makes it easy to compute the exchange-correlation potential,

$$v_{xc}[\rho(r)] \cong \frac{de_{xc}[\rho(r)]}{d\rho(r)} \equiv \mu_{xc}[\rho(r)], \tag{20}$$

where $\mu_{xc}[\rho(r)]$ is the exchange-correlation chemical potential of the uniform electron gas with density equal to the local density $\rho(r)$.

A particularly simple, albeit somewhat inaccurate, form for $e_{xc}[\rho(r)]$ is obtained by neglecting correlation altogether in the uniform gas. This produces the *exchange-only* form[20]

$$e_{xc}[\rho(r)] \cong -\frac{3}{4}\left(\frac{3\rho(r)}{\pi}\right)^{\frac{1}{3}}\rho(r). \tag{21}$$

and the potential

$$v_{xc}[\rho(r)] \cong -\left(\frac{3\rho(r)}{\pi}\right)^{\frac{1}{3}}, \tag{22}$$

both in atomic units.

Various forms incorporating correlation have been proposed, and produce significant improvement over the exchange-only approximation, as illustrated by the numerical results for the ground-state energies of He, Ne, and Ar, in Table 3.. A particularly accurate expression for the exchange-correlation energy has been proposed by Vosko *et al.*[21] on the basis of a Monte-Carlo computation by Ceperley and Alder[22].

Most of the discrepancy between the calculated and experimental results in Table 3. is due to self interaction (*i.e.*, the interaction of an electron with its own potential). This interaction, introduced by the second term on the right-hand side of Eq. (17), is expected to be cancelled out by a contribution of the opposite sign in the exchange-correlation potential. Unfortunately, the

Table I: LDA exchange-only[23] and exchange-correlation [8] results for three rare-gas-atom ground-state energies.

Atom	Energy (Ry) Exchange	Exchange-correlation	Experimental
He	-5.447	-5.651	-5.807
Ne	-254.986	-256.349	-257.880
Ar	-1049.027	-1051.709	-1055.21

LDA fails to produce this negative contribution, so that local-density computations tend to overestimate the ground-state energies of localized systems. The self-interaction corrections to the LDA proposed by Perdew and Zunger[24] avoid this problem, make the energies produced by the exchange-only approximation agree very well with the Hartree-Fock results and those produced by better exchange-correlation forms agree very well with the experimental results. This, however, is a subject outside the scope of the present review.

4. RAYLEIGH-RITZ PRINCIPLE FOR EXCITED STATES

The Rayleigh-Ritz principle supporting the Hohenberg-Kohn theorem in Section 5. concerns the spectrum $E_1 \leq E_2 \leq \ldots$ of a Hamiltonian $\hat{H}$ with eigenstates $|j\rangle$, $(j = 1, 2, \ldots)$. To state the principle, two definitions are necessary: for given integer m, the set $\mathcal{L}_m$ is defined as the subspace spanned by all the eigenstates $|j\rangle$ with $E_j < E_m$, and $\mathcal{U}_m$ as the subspace spanned by all the eigenstates $|j'\rangle$ with $E_{j'} \leq E_m$. As example, consider a spectrum with $E_1 < E_2 = E_3 = E_4 < E_5 = E_6 < E_7$; for $m = 5$ (or $m = 6$), the following structure defines the two subspaces:

$$\begin{array}{ll} & \left.\begin{array}{l} E_5 = E_6 \\ \\ E_2 = E_3 = E_4 \\ \\ E_1 \end{array}\right\} \mathcal{U}_m. \\ \mathcal{L}_m & \left\{\begin{array}{l} E_2 = E_3 = E_4 \\ E_1 \end{array}\right. \end{array}$$

In this notation, here is the Rayleigh-Ritz principle:

(a) *Let $w_1 \geq w_2 \geq \ldots \geq w_M$ be real positive numbers. The following inequality is then satisfied for any set of orthonormal functions $\{|\phi_1\rangle, |\phi_2\rangle, \ldots, |\phi_M\rangle\}$:*

$$w_1 \langle \phi_1 | H | \phi_1 \rangle + \cdots + \langle \phi_M | H | \phi_M \rangle \geq w_1 E_1 + \cdots + w_M E_M. \tag{23}$$

(b) *The equality in (23) holds if and only if for $m = M$, and for all other m with $w_m \neq w_{m+1}$, the subspace $[|\phi_1\rangle, \ldots, |\phi_m\rangle]$, spanned by the trial functions $|\phi_1\rangle, \ldots, |\phi_m\rangle$, lies in $\mathcal{U}_m$ and contains $\mathcal{L}_m$,* i.e.,

$$\mathcal{L}_m \subset [|\phi_1\rangle, \ldots, |\phi_m\rangle] \subset \mathcal{U}_m. \tag{24}$$

The second part of the theorem is sufficiently complex to justify an illustrative example. Accordingly, instead of reproducing the general proof of the theorem, the following discussion is particularized to the spectrum in the scheme above and to ensembles containing $M = 2$ states.

In this case, two trial functions $|\phi_1\rangle$ and $|\phi_2\rangle$ have to be considered, each a linear combination of the eigenstates of H:

$$|\phi_1\rangle = \sum_j \alpha_j |j\rangle, \tag{25}$$

and

$$|\phi_2\rangle = \sum_j \beta_j |j\rangle. \tag{26}$$

Here, since $|\phi_1\rangle$ and $|\phi_2\rangle$ are orthonormal, the linear coefficients α_j and β_j satisfy

$$\sum_j |\alpha_j|^2 = \sum_j |\beta_j|^2 = 1 \tag{27}$$

and

$$|\alpha_j|^2 + |\beta_j|^2 \leq 1 \qquad (j = 1,2,\ldots). \tag{28}$$

In view of Eqs. (25) and (26), the difference between the right-hand and left-hand sides of Eq. (23) can be written

$$\begin{aligned} w_1(\langle\phi_1|H|\phi_1\rangle - E_1) + w_2(\langle\phi_2|H|\phi_2\rangle) - E_2) &= \\ w_1 \sum_{j>1} |\alpha_j|^2(E_j - E_1) + &\qquad w_2 \sum_{j>2} |\beta_j|^2(E_j - E_2) \end{aligned} \tag{29}$$

equivalent to

$$\begin{aligned} w_1(\langle\phi_1|H|\phi_1\rangle - E_1) + w_2(\langle\phi_2|H|\phi_2\rangle) - E_2) &= \\ w_1 \sum_{j>2} |\alpha_j|^2(E_j - E_2) + &\qquad w_2 \sum_{j>2} |\beta_j|^2(E_j - E_2) \\ +w_1 \sum_{j>1} |\alpha_j|^2(E_2 - E_1) + &\qquad w_2|\beta_1|^2(E_1 - E_2). \end{aligned} \tag{30}$$

The terms proportional to $E_2 - E_1$ on the right-hand side can now be combined, which yields

$$\begin{aligned} w_1(\langle\phi_1|H|\phi_1\rangle - E_1) + w_2(\langle\phi_2|H|\phi_2\rangle) - E_2) &= \\ w_1 \sum_{j>2} |\alpha_j|^2(E_j - E_2) + &\qquad w_2 \sum_{j>2} |\beta_j|^2(E_j - E_2) \\ +w_1(1 - |\alpha_1|^2) - &\qquad w_2|\beta_1|^2](E_2 - E_1). \end{aligned} \tag{31}$$

Finally, it is convenient to add and subtract w_2 to the weight w_1 in the square brackets on the right-hand side, so that Eq. (32) becomes

$$\begin{aligned} w_1(\langle\phi_1|H|\phi_1\rangle - E_1) + w_2(\langle\phi_2|H|\phi_2\rangle) - E_2) &= \\ w_1 \sum_{j>2} |\alpha_j|^2(E_j - E_2) + &\qquad w_2 \sum_{j>2} |\beta_j|^2(E_j - E_2) \\ +(w_1 - w_2)(1 - |\alpha_1|^2) + &\qquad w_2(1 - |\alpha_1|^2 - |\beta_1|^2)(E_2 - E_1). \end{aligned} \tag{32}$$

For $M = 2$, this inequality proves the teorem. Part (a) follows from observing that, since $E_j \geq E_i$ $(j > i)$, since $w_2 \geq w_1$, and $1 - |\alpha_1|^2 - |\beta_1|^2 \geq 0$ [see Eq. (28)], each term on the right-hand side is non-negative. Part (b) requires that we examine the conditions under which each term vanishes. As an example, we consider the spectrum drawn above and assume that $w_1 > w_2$. Since $E_5 > E_4 = E_3 = E_2$, the first term on the right-hand side equals zero if and only if $\alpha_j = 0$ for $j > 3$; according to Eq. (25), this is equivalent to

$$[|\phi_1\rangle, |\phi_2\rangle] \subset [|1\rangle, |2\rangle, |3\rangle |4\rangle]. \tag{33}$$

Since $E_2 > E_1$, on the other hand, the last term on the right-hand side of Eq. (33) will vanish if and only if $\alpha_1 = 1$ and $\beta_1 = 0$, equivalent to

$$|\phi_1\rangle = |1\rangle. \tag{34}$$

Eqs. (33) and (34) are precisely the conditions stipulated by Eq. (24). Since $w_1 \neq w_2$, that relation must be analyzed for $m = 1$ and for $m = 2$. In the first case, $\mathcal{L}_{m=1} = \{\}$ and $\mathcal{U}_{m=1} = \{|1\rangle\}$, so that Eq. (24) becomes

$$\{|\phi_1\rangle\} = \{|1\rangle\}, \tag{35}$$

i.e., $|\phi_1\rangle = |1\rangle$.

For $m = 2$, on the other hand, since $\mathcal{L}_{m=2} = \{|1\rangle\}$, and $\mathcal{U}_{m=2} = [|1\rangle, |2\rangle, |3\rangle]$, Eq. (24) becomes

$$\{|1\rangle\} \subset [|\phi_1\rangle, |\phi_2\rangle] \subset [|1\rangle, |2\rangle, |3\rangle], \tag{36}$$

equivalent to Eqs. (33) and (34).

It should be clear from this discussion that, while part (a) of the theorem is a straightforward extension of the Rayleigh-Ritz principle for equiensembles, part (b) is not. The two segments are nevertheless equally important for the Hohenberg-Kohn theorem for ensembles of fractionally occupied states, as Section 5. will show.

5. DENSITY FUNCTIONAL FORMALISM

5.1. The Hohenberg-Kohn Theorem

The variational principle in Section 4. extends the Hohenberg-Kohn theorem to equiensembles as well as to ensembles of fractionally occupied states. A proof of the theorem based on the constrained search formalism of Levy and Lieb[16,17] exists[4], but this chapter, which ignores the v-representability problem, will follow the approach of Hohenberg-Kohn paper[1].

The set $W = \{w_1, \ldots, w_M\}$ defines the ensemble density $\rho^W = w_1\langle 1|\psi^\dagger\psi|1\rangle + \cdots + w_M\langle M|\psi^\dagger\psi|M\rangle$. The Hohenberg-Kohn theorem states that this density determines the external potential $v(r)$ in the Hamiltonian

$$\begin{aligned} \hat{H} &= \int \psi^\dagger(r)\frac{-\nabla^2}{2}\psi(r)\,dr + \int \psi^\dagger(r)v(r)\psi(r)\,dr \\ &+\frac{1}{2}\int \psi^\dagger(r)\psi^\dagger(r')\frac{1}{|r-r'|}\psi(r')\psi(r)\,dr\,dr', \end{aligned} \tag{37}$$

up to within an additive constant. Following the *reductio ad absurdum* reasoning of Hohenberg and Kohn, we assume that the Schrödinger equations for two potentials $v(r)$ and $v'(r)$, differing by more than a constant, have eigenstates $|1\rangle, \ldots, |M\rangle$, and $|1'\rangle, \ldots, |M'\rangle$ that correspond to identical equiensemble densities $\rho^W(r) = \rho'(r)$, respectively[25].

Referring to the Rayleigh-Ritz principle in Section 4., we now take the eigenstates $|1'\rangle, \ldots, |M'\rangle$ of $\hat{H}' = \hat{T} + \hat{U} + \hat{V}'$ as trial states for the Hamiltonian $\hat{H} = \hat{T}+\hat{U}+\hat{V}$. As shown in[4], these trial states fail to satisfy condition (24), in part (b) of the Rayleigh-Ritz principle. The *strict* inequality therefore holds in (23), *i.e.,*

$$w_1\langle 1|H|1\rangle + \cdots + w_M\langle M|H|M\rangle < w_1\langle 1'|H|1'\rangle + \cdots + w_M\langle M'|H|M'\rangle. \tag{38}$$

$\hat{H}' - \hat{V}' + \hat{V}$ substituted for H on the right-hand side, we find

$$\begin{aligned} w_1\langle 1|H|1\rangle + \cdots + w_M\langle M|H|M\rangle &< w_1\langle 1'|H'|1'\rangle + \cdots + w_M\langle M'|H'|M'\rangle \\ &+\int \rho^W(r')[v(r) - v'(r)]\,dr. \end{aligned} \tag{39}$$

Precisely the same reasoning can now be followed with $|1\rangle, \ldots, |M\rangle$ as trial states for the Hamiltonian $\hat{H}'$. We find

$$\begin{aligned} w_1\langle 1'|H'|1'\rangle + \cdots + w_M\langle M'|H'|M'\rangle &< w_1\langle 1|H|1\rangle + \cdots + w_M\langle M|H|M\rangle \\ &+\int \rho^W(r')[v'(r) - v(r)]\,dr. \end{aligned} \tag{40}$$

Added, Eqs. (39) and (40) produce the inconsistency $0 < 0$, contradicting the assumption that $|1\rangle, \ldots, |M\rangle$ and $|1'\rangle, \ldots, |M'\rangle$ have the same ensemble density. This shows that, apart from an additive constant, the ensemble density $\rho'(r)$ determines uniquely the potential $v'(r)$. For non-degenerate spectra, it follows that $\rho'(r)$ determines the eigenstates $|1'\rangle, \ldots, |M'\rangle$, so that the ensemble-averaged expectation value

$$E_v^W = w_1\langle 1'|H|1'\rangle + \ldots + w_M\langle M'|H|M'\rangle, \tag{41}$$

which, for $\rho'(r) = \rho^W(r)$, reduces to the ensemble energy

$$\mathcal{E}^W = w_1 E_1 + \ldots + w_M E_M, \tag{42}$$

is a functional $E_v^W[\rho']$ of the ensemble density. Moreover, for any density $\rho'(r) \neq \rho^W(r)$, the Rayleigh-Ritz principle for excited states guarantees the *strict* inequality

$$\mathcal{E}^W = E_v^W[\rho^W] < E_v^W[\rho']. \tag{43}$$

For degenerate spectra, the potential $v'(r)$ does not determine uniquely the eigenstates $|1'\rangle, \ldots, |M'\rangle$. Nonetheless, as[4] shows, the ensemble-averaged expectation value in Eq. (41) is a (unique) functional of the ensemble density and satisfies the strict inequality in (43).

5.2. The Kohn-Sham Equations

The variational principle (43) leads naturally to Kohn-Sham-like relations determining the equiensemble density $\rho^W(r)$. To generate these single-particle equations, we have to consider a non-interacting electron gas in a potential $v_s[\rho^W](r)$ such that the ensemble density for the non-interacting system equals $\rho^W(r)$. If v_s were known, the single-particle wave-function $\varphi_j(r)$ and energies ϵ_j for the jth orbital of that system could be easily obtained from the Schrödinger equation

$$[-\frac{\nabla^2}{2} + v_s(r)]\varphi_j(r) = \epsilon_j\varphi_j(r), \tag{44}$$

and the ensemble density could be subsequently calculated:

$$\rho^W(r) = \sum_{m=1}^{M} w_m \sum_{j=1}^{\infty} f_{jm} |\varphi_j(r)|^2, \tag{45}$$

where the occupation f_{jm} is unity (zero) if the jth orbital is filled (empty) in the mth state of the non-interacting ensemble.

As in the ground-state formalism, the potential v_s is determined variationally. An (approximate) expression for $E_v^W[\rho']$ being therefore necessary, it is convenient to define the ensemble exchange-correlation energy functional $E_{xc}^W[\rho']$. We therefore write

$$\begin{aligned} E_v^W[\rho'] &= T_s[\rho'] + \frac{1}{2}\int \frac{\rho'(r)\rho'(r')}{|r-r'|}\,dr\,dr' \\ &\quad + \int \rho'(r)v(r)\,dr + E_{xc}^W[\rho']. \end{aligned} \tag{46}$$

Here, $T_s[\rho']$ is the kinetic energy of the non-interacting system:

$$T_s[\rho'] = \sum_{m=1}^{M} w_m \sum_j f_{jm} \int \varphi_j^*(r)(-\frac{\nabla^2}{2})\varphi_j(r)\,dr. \tag{47}$$

To determine this functional of the density $\rho'(r)$, we multiply both sides of Eq. (44) by $\varphi^*(r)$, integrate them over r, and sum them over the occupied orbitals in each state m. Eq. (47) then yields

$$T_s[\rho'] = \sum_{m=1}^{M} w_m \sum_j f_{jm}\varepsilon_j - \int \rho'(r)v_s(r)\,dr, \tag{48}$$

and Eq. (46) becomes

$$\begin{aligned} E_v^W[\rho'] &= \sum_{m=1}^{M} w_m \sum_j f_{jm}\varepsilon_j - \int \rho'(r)[v_s(r) - v(r)]\,dr \\ &\quad + \frac{1}{2}\int \frac{\rho'(r)\rho'(r')}{|r-r'|}\,dr\,dr' + E_{xc}^W[\rho']. \end{aligned} \tag{49}$$

Applied to this equation, the condition $\delta E_v^W/\delta\rho^W = 0$ now determines the potential $v_s[\rho^W](r)$. From perturbation theory, the first order variation $\delta\varepsilon_j$ of the eigenvalue in (44) due to a change $\delta\rho'$ in the ensemble density is $\delta\varepsilon_j = \int |\varphi_j(r)|^2 \delta v_s(r)\,dr$, and hence

$$\sum_{m=1}^{M} w_m \sum_j f_{jm}\delta\varepsilon_j = \rho^W(r)\delta v_s(r). \tag{50}$$

The variations of the other terms on the right-hand side of Eq. (49) are easily found, leading to

$$v_s[\rho^W](r) = v(r) + \int \frac{\rho^W(r')}{|r - r'|}\, dr' + v_{xc}^W[\rho^W](r) \tag{51}$$

where, as in the ground-state formalism,

$$v_{xc}^W[\rho^W](r) = \delta E_{xc}^W[\rho^W]/\delta\rho^W(r). \tag{52}$$

Eq. (51) completes the cycle of Kohn-Sham equations that (44) and (45) started. Substitution of (51) for $v_s(r)$ on the right-hand side of Eq. (49) yields a relation similar to the Kohn-Sham expression for the ground-state energy:

$$\begin{aligned} \mathcal{E}^W = E_v^W[\rho^W] \;&=\; \sum_{m=1}^{M} w_m \sum_j f_{jm}\varepsilon_j - \frac{1}{2}\int \frac{\rho^W(r)\rho^W(r')}{|r - r'|}\, dr\, dr' \\ &\quad - \int \rho^W(r) v_{xc}[\rho^W](r)\, dr + E_{xc}^W[\rho^W]. \end{aligned} \tag{53}$$

5.3. Calculation of Excitation Energies

5.3.1. The equiensemble approach. Eq. (53) determines the ensemble-averaged energies $\mathcal{E}^W$, linear combinations of the eigenvalues E_j $(j = 1, \ldots, M)$ from which the latter energies can be recovered. As an example, consider the equiensemble, whose identical weights,

$$w_1 = \ldots = w_M = 1/M, \tag{54}$$

suggest a more specific notation for the ensemble-averaged energy, $\mathcal{E}_M$ instead of $\mathcal{E}^W$. In this case, the following relation, equivalent to Eq. (42), determines the energy of the g-fold degenerate multiplet encompassing the eigenstates $|M - g + 1\rangle, \ldots, |M\rangle$:

$$E_M = \mathcal{E}_M + \frac{M - g}{g}(\mathcal{E}_M - \mathcal{E}_{M-g}). \tag{55}$$

To compute the I lowest energies in a spectrum with degeneracies $g_1, \ldots, g_I$, one then has to calculate $\mathcal{E}_M$ for $M =$

$g_1, \ldots, g_1 + \cdots + g_I$. This simple procedure defines the equiensemble approach.

5.3.2. The method of fractional occupation. Inspection of Eq. (55) reveals a shortcoming of the equiensemble method. The expression within parentheses on the right-hand side is a small number obtained by subtracting two large equiensemble energies, associated with two different densities and two different M. The subtraction can therefore introduce large relative errors. This section discusses an alternative procedure (the fractional occupation approach) that avoids this problem.

As Section 6. will show, the currently available expression for the exchange-correlation functional $E_{xc}^W[\rho^W]$ describes only the equiensemble. This circumstance makes Eq. (55) (involving only equiensemble energies) very convenient and suggests that, to calculate excited state energies, we employ either the weights (54) (the equiensemble approach in Section 5.3.1.) or weights that reduce to the equiensemble in appropriate limits, a possibility the following definition explores:

$$\begin{cases} w_1 = \ldots = w_{M-g} = (1 - wg)/(M - g) \\ w_{M-g+1} = \ldots = w_M = w \end{cases} \qquad (0 \leq w \leq \frac{1}{M}), \qquad (56)$$

where w is a free parameter. For $w = 1/M$, these weights reduce to $1/M$, and we recover Eq. (54). For $w = 0$, the g-fold degenerate multiplet $|M - g + 1\rangle, \ldots, |M\rangle$ has zero weight, and the others have weights $w_1 = \ldots = w_{M-g} = 1/(M - g)$, so that we again recover Eq. (54), with $M \to M - g$. Eq. (56) hence interpolates between equiensembles. The added flexibility supplied by the parameter w opens a new avenue leading to excitation energies, as the following discussion shows.

With the weights (56) substituted on the right-hand side, Eq. (42) can be differentiated with respect to w. We find

$$\frac{d\mathcal{E}^W}{dw} = gE_M - \frac{g}{M-g} \sum_{m=1}^{M-g} E_m \qquad (57)$$

and, after substituting Eq. (55) for E_M on the right-hand side,

$$\frac{d\mathcal{E}^W}{dw} = M(\mathcal{E}_M - \mathcal{E}_{M-g}). \tag{58}$$

Now, $\mathcal{E}^W$ is given by Eq. (53), which with the present weights reads

$$\begin{aligned} E_v^W[\rho^W] &= \frac{1-wg}{M-g}\sum_{m=1}^{M-g}\sum_j f_{jm}\varepsilon_j + w\sum_{m=M-g+1}^{M}\sum_j f_{jm}\varepsilon_j + E_{xc}^W[\rho^W] \\ &\quad - \int \frac{\rho^W(r)\rho^W(r')}{2|r-r'|}\,dr\,dr' - \int \rho^W(r)v_{xc}[\rho^W](r)\,dr. \end{aligned}$$

By differentiating this equality with respect to w we can therefore calculate the left-hand side of Eq. (58). Although the ensemble density $\rho^W(r)$ (produced by the Kohn-Sham equations) depends on w, this derivative is very easily computed, because the variational property of $E_{xc}^W[\rho^W]$ ensures that $dE_v^W[\rho^W]/dw = \partial E_v^W[\rho^W]/\partial w$. It follows that

$$\begin{aligned} \frac{dE_v^W[\rho^W]}{dw} &= \sum_{m=M-g+1}^{M}\sum_j f_{jm}\varepsilon_j - \frac{g}{M-g}\sum_{m=1}^{M-g}\sum_j f_{jm}\varepsilon_j \\ &\quad + \left.\frac{\partial E_{xc}^W[\rho^W]}{\partial w}\right|_{\rho^W(r)}, \end{aligned} \tag{59}$$

and from Eq. (58) that

$$\begin{aligned} M(\mathcal{E}_M - \mathcal{E}_{M-g}) &= \sum_{m=M-g+1}^{M}\sum_j f_{jm}\varepsilon_j - \frac{g}{M-g}\sum_{m=1}^{M-g}\sum_j f_{jm}\varepsilon_j \\ &\quad + \left.\frac{\partial E_{xc}^W[\rho^W]}{\partial w}\right|_{\rho^W(r)}. \end{aligned} \tag{60}$$

To compute the energy difference on the right-hand side of (55) it is hence sufficient to solve the Kohn-Sham equations for a single w in the interval $0 \le w \le 1/M$, so that (60) avoids the numerical problem in the equiensemble approach. Practical applications of this equality of course require an approximation for $\partial E_{xc}^W[\rho^W]/\partial w$. For $w = 1/2M$, Section 6. presents a simple expression for this derivative.

6. THE QUASI-LOCAL DENSITY APPROXIMATION

6.1. The Equiensemble Exchange-Correlation Functionals

This section discusses the QLDA for the ensemble-averaged exchange-correlation energy functional[3]. This approximation replaces ensemble averages (for which, as Section 3. pointed out, the LDA is inadequate) by thermal averages (for which the LDA is appropriate). It is therefore expected to become accurate for $M \to \infty$, the limit in which the thermal ensemble and the equiensemble become identical.

In the Kohn-Sham scheme of Section 5.2., *two* ensembles of identical multiplicity M are involved, one of interacting and the other of non-interacting particles. With them we will associate two canonical ensembles, at temperatures θ and θ_s, respectively. Each equiensemble has entropy $k_B \ln M$, where k_B is Boltzmann's constant. Therefore, the following LDA fixes the two temperatures:

$$\int \sigma^{\theta}[\rho^W](r)\,dr = \int \sigma_s^{\theta_s}[\rho^W](r)\,dr = k_B \ln M. \tag{61}$$

Here and henceforth, the symbols $[\rho](r)$ denote the LDA, indicating the preceding quantity [in this case σ^{θ} ($\sigma_s^{\theta_s}$), the entropy per unit volume of interacting (non-interacting) electrons at temperature θ (θ_s)] for a homogeneous system with the local density $\rho(r)$.

Eq. (46) expresses the exchange-correlation energy functional in terms of the equiensemble energy $E_v^W[\rho']$ and of the kinetic energy $T_s^W[\rho']$. For large M, the former functional is approximately equal to the canonical-ensemble energy functional $E_v^{\theta}[\rho']$, in turn related to the canonical-ensemble exchange-correlation energy functional $E_{xc}^{\theta}[\rho']$ by the thermal analogue of (46)[5]:

$$E_v^{\theta}[\rho'] = T_s^{\theta}[\rho'] + \int \frac{\rho'(r)\rho'(r')}{2|r-r'|}\,dr\,dr' + \int \rho'(r)v(r)\,dr + E_{xc}^{\theta}[\rho']. \tag{62}$$

The right-hand side of this equation substituted for $E_v^W[\rho']$ and the canonical-ensemble kinetic-energy functional at temperature θ_s substituted for $T_s^W[\rho']$, Eq. (46) becomes

$$E_{xc}^W[\rho'] = T_s^{\theta}[\rho'] + E_{xc}^{\theta}[\rho'] - T_s^{\theta_s}[\rho'], \tag{63}$$

thus relating the equiensemble exchange-correlation energy functional to canonical-ensemble functionals, for which the LDA is appropriate. For systems of slowly varying density, therefore,

$$E_{xc}^W[\rho'] \approx \int \{t_s^{\theta}[\rho'](r) + e_{xc}^{\theta}[\rho'](r) - t_s^{\theta_s}[\rho'](r)\}\, dr, \tag{64}$$

where t_s^{θ} (e_{xc}^{θ}) denotes the kinetic (exchange-correlation) energy per unit volume for a homogeneous non-interacting (interacting) system at temperature θ.

Eqs. (52) and (64) determine the equiensemble exchange-correlation potential. Since W represents the equiensemble, the derivative on the right-hand side of (52) must be taken at *fixed entropy*. With the thermodynamical identity $\mu = \partial E(\rho, S)/\partial\rho|_S$ (relating internal energy E, entropy S, and chemical potential μ), simple manipulations lead to

$$v_{xc}^W[\rho'](r) \approx \mu_s^{\theta}[\rho'](r) + \mu_{xc}^{\theta}[\rho'](r) - \mu_s^{\theta_s}[\rho'](r), \tag{65}$$

where μ_s^{θ} (μ_{xc}^{θ}) denotes the chemical potential (the exchange-correlation chemical potential) for homogeneous non-interacting (interacting) electrons at temperature θ.

This approximation for the exchange-correlation potential and Eq. (64) for the exchange-correlation energy functional complement the Kohn-Sham equations for the equiensemble approach. The two expressions, weakly non-local because the temperatures θ and θ_s are non-local functionals of the density [see (61)], are made practical by satisfactorily accurate approximations for the various thermal properties of the homogeneous interacting and non-interacting electron gases appearing on the right-hand sides[11].

For $M = 1$ ($\theta = \theta_s = 0$), Eqs. (64) and (65) reduce to the

LDA for the ground-state exchange-correlation energy functional and potential, respectively. The QLDA is therefore accurate both for $M \to 1$ and for $M \to \infty$; it should be satisfactory even for moderately large M. The numerical results in Section 7. support this conclusion.

6.2. Application to the Fractional Occupation Method

6.2.1. Approximate form for $\partial E_{xc}^W[\rho^W]/\partial \rho^W(r)$. As we have just seen, the QLDA provides an expression for the equiensemble exchange-correlation functional, hence making practical the equi-ensemble approach. In addition, as the present section shows, it produces an approximation for the last term on the right-hand side of Eq. (60), hence making practical the fractional occupation approach.

Section 5.3.2. has pointed out that, for $w = 0$ and $1/M$, the weights w_m in Eq. (56) reduce to the equiensembles with multiplicities $M - g$ and M, respectively. To these the QLDA applies, suggesting the following expression for the partial derivative in (60):

$$\left.\frac{\partial E_{xc}^W[\rho^W]}{\partial w}\right|_{\rho^W} = M\{E_{xc}^{w=1/M}[\rho^{w=1/2M}] - E_{xc}^{w=0}[\rho^{w=1/2M}]\} + \mathcal{O}(1/M^2). \qquad (66)$$

Although the right-hand side involves the energy functionals for $w = 0$ and $1/M$, *i.e.*, $E_{xc}^{M-g}[\rho']$ and $E_{xc}^{M}[\rho']$, respectively, both functionals are evaluated at the density $\rho^{w=1/2M}(r)$, *i.e.*, at the ensemble density found by solving the Kohn-Sham equations in Section 5.2. with $w = 1/2M$. This weight eliminates the term of $\mathcal{O}(1/M)$ from (66).

To compute $\rho^{w=1/2M}$ we also need the exchange-correlation potential $v_{xc}^{w=1/2M}$. In analogy with Eq. (66), we can write

$$v_{xc}^{w=1/2}[\rho^W](r) = \frac{v_{xc}^{w=0}[\rho^W](r) + v_{xc}^{w=1/M}[\rho^W](r)}{2} + \mathcal{O}(1/M^2). \qquad (67)$$

With $v_{xc}^{w=0}$ and $v_{xc}^{w=1/M}$ given by (65), Eqs. (66) and (67) turn the fractional occupation approach into a practical computational procedure. An illustration will be presented in Section 7.. Before that, however, the next section discusses an important alternative.

6.2.2. Comparison with the transition state approach. Consider an N-electron interacting system with non-degenerate ground state and g-fold degenerate first excited state. Applied to the ensemble multiplicity $M = 1 + g$, (60) gives its first excitation energy:

$$\overline{E}_2 - E_1 = \varepsilon_{N+1} - \varepsilon_N + \frac{1}{g} \left. \frac{\partial E_{xc}^W[\rho^W]}{\partial w} \right|_{\rho^W(r)}, \tag{68}$$

where $\overline{E}_2 = E_2 = \ldots = E_M$ is the second-multiplet energy. ε_N (ε_{N+1}) is the Nth ($N+1$th) Kohn-Sham eigenvalue, *i.e.*, the last occupied (first vacant) level in the non-interacting ground state. Both correspond to the ensemble density $\rho^W = (1 - wg)\rho_1 + wg\overline{\rho}_2$, where ρ_1 ($\overline{\rho}_2$) is the ground-state (average first-excited-multiplet) density. Compare this equation with the transition state formula[15]:

$$\overline{E}_2 - E_1 = \varepsilon^S_{N+1} - \varepsilon^S_N, \tag{69}$$

where ε^S_N denotes the Nth eigenvalue of the Hartree-Fock-Slater equations for the density $\rho^S = (\rho_1 + \overline{\rho}_2)/2$. Three distinctions separate Eqs. (68) and (69): (1) the Hartree-Fock-Slater equations take only ground-state exchange into account, whereas the Kohn-Sham like equations involve the ensemble averaged exchange-correlation energy functional; (2) for $w \neq 1/2g$, the densities ρ^S and ρ^W are different, and (3) the last term on the right-hand side of Eq. (68) is missing from Eq. (69). Nevertheless, for $w = 1/2M \equiv 1/2(g+1)$, Eq. (66) provides an estimate of that term; as it turns out, its contribution to Eq. (68) is small[26]. Thus, for highly degenerate first excited state (*i.e.*, for $M \approx g$), Eqs. (68) and (69) yield approximately equal results.

From this, one might conclude that the fractional occupation method is a refinement of the transition state method. This inference cannot be generalized, however: to calculate the Ith energy in the spectrum, the transition state approach prescribes $\rho^S = (\rho_1 + \overline{\rho}_I)/2$, while ρ^W is the ensemble average of the densities $\rho_1, \overline{\rho}_2, \ldots, \overline{\rho}_I$. Moreover, for $I > 2$, (60) is very different from the equivalent of (69)[15]. For energies higher than $\overline{E}_2$, therefore, the two methods can be compared only on the basis of the numerical results they produce. This considered, a more conclusive comparison will be deferred to Section 7.2..

Like the quasi-local-density approximation, the approximations implicit in Eqs. (66) and (67) become accurate for large M. To test them for small M, one must compare equiensemble energies computed through the fractional occupation approach [which employs Eqs. (66) and (67)] with the same energies found through the equiensemble approach (which does not employ them). One such test, to be reported in Section 7., indicates that the neglected terms [*i.e.* the terms of $\mathcal{O}(1/M^2)$ and of $\mathcal{O}(1/M)$] affect insignificantly the calculated excitation energies. Even for small M the finite differences approximations seem satisfactory.

7. NUMERICAL RESULTS FOR THE He ATOM

7.1. Equiensemble Energies

The He atom, a small system with widely-studied excitation spectrum and negligible spin-orbit interaction[27], is a convenient testground for the approximations in Section 6.. The numerical computation[26] of its equiensemble energies by the two methods in Section 5.3. will be reported in this section. To recapitulate, in the equiensemble procedure, Eq. (53) yields the ensemble-averaged energies. In the fractional occupation approach, Eq. (60) determines equiensemble-energy differences; from these, the equiensemble energies measured from a reference level, such as the ground-state energy, can be computed.

Those expressions make no reference to electron spin; ensemble-density-functional theory has so far been limited to spin-averaged densities. This restricts the following discussion to Helium densities and energies averaged over the *para* ($S = 0$) and *ortho* ($S = 1$) configurations of each excited state, weights one and three, respectively.

The energies are then labeled by principal quantum number n and angular momentum L. Each multiplet nL has degeneracy $g = 4(2L+1)$. In increasing-energy order, the degeneracies $g = 1$ [the 1S (ground) state], $g = 4$ (2S), $g = 12$ (2P), ..., $g = 4$ (5S) form the ensemble multiplicities $M = 1, 5, 17, \ldots, 121$. These and the ensemble-averaged experimental energies (reckoned from the experimental ionization threshold, $E_1^{+EXPT} = -4\,\mathrm{Ry}$) appear in Table II. The equiensemble energies [measured from the ground-state energy ($E_1^{+LDA} = -3.725\,\mathrm{Ry}$) for the ionized He atom, calculated in the LDA[21]] are also tabulated.

Table II: Equiensemble energies[a] and energy differences in the equiensemble (EQ) and fractional occupation (FO) approaches, compared with the experimental (EXPT) values.

M	Multiplet[b]	$\mathcal{E}_M^{EXPT}$	$\mathcal{E}_M^{EQ}$	$\Delta\mathcal{E}_M^{EQ}$	$\Delta\mathcal{E}_M^{FO}$	$\Delta\mathcal{E}_M^{EXPT}$
1	1S	-1.807	-1.947(8%)			
5	2S	-0.630	-0.727(15%)	1.220	1.240	1.177
17	2P	-0.370	-0.386(4%)	0.341	0.346	0.260
21	3S	-0.325	-0.345(6%)	0.041	0.041	0.045
33	3P	-0.249	-0.265(6%)	0.081	0.083	0.077
53	3D	-0.197	-0.200(2%)	0.065	0.067	0.052
57	4S	-0.188	-0.192(2%)	0.008	0.008	0.009
69	4P	-0.166	-0.171(3%)	0.020	0.020	0.022
89	4D	-0.143	-0.146(2%)	0.025	0.024	0.023
117	4F	-0.124	-0.123(1%)	0.024	0.023	0.019
121	5S	-0.121	-0.120(1%)	0.002	0.002	0.003

[a] Energies (Ry) measured from the ionization threshold. Percentages show deviations from experimental values.
[b] Highest multiplet in equiensemble.

Subtracting E_1^+ from each set of equiensemble energies eliminates from Table II the unphysical self-interactions contained in the energies calculated with the LDA[24] (for example, most of the discrepancy between E_1^{+EXPT} and E_1^{+LDA} is self-interaction) and, by extension, in those calculated with the QLDA. This focuses attention on specific properties of the latter approximation.

The agreement between the calculated and experimental equiensemble energies goes from good to excellent as M increases. This is of course expected, the QLDA being accurate for large M. Although M-dependent, the deviation between the calculated and the experimental $\mathcal{E}_M$ follows a pattern: for growing M, it is significantly reduced whenever a new angular momentum is introduced. The deviation, nearly 10% for the (two) equiensembles involving only S states, decreases to 5% for those involving S and P states, to 2% for S, P, and D states, and to 1% for S, P, D, and F states. This seems to be due to the LDA embedded in the QLDA, for an ensemble rich in angular momenta should be well represented by a homogeneous electron gas.

This non-uniform convergence of the calculated equiensemble energies to the experimental values affects perniciously the equiensemble-energy differences; Eq. (55) shows that the calculated excited-state energies will be likewise degraded. In light of the introductory argument in Section 5.3.2., one might expect Eq. (60) to yield more accurate equiensemble-energy differences. Table II indicates that it does not, apparently because the approximation (66) for $\partial E_{xc}^{W}[\rho^{W}]/\partial\rho^{W}(r)$ inherits the limitations of the QLDA. In particular, the last two terms in that equation involve ensemble multiplicities M and $M-g$, respectively, so that the fluctuating accuracy of the QLDA defiles the equiensemble-energy differences calculated with the fractional occupation method just as it corrodes precision in the equiensemble method. From this standpoint, the agreement between the equiensemble energy differences calculated by the two procedures is disappointing.

7.2. Excitation Spectrum

Table III displays the multiplet energies $\overline{E}_I$, $(I = 1,\ldots,11)$, derived from the equiensemble energies and energy differences in Table II. All levels have been subtracted from the ionization threshold. Since the fractional occupation approach produces only equiensemble energy differences, to compute the energy in Eq. (55) it is necessary to measure them from a reference level. The calculation becoming more accurate for large M, it is convenient to reckon the equiensemble energies from the the largest experimental energy in the spectrum, *i.e.*, in Table III, from E_{11}^{EXPT}. In practice, this makes $E_{11}^{FO} = E_{11}^{EXPT}$.

As already discussed in Section 7.1., the agreement between the energies produced by the two approaches is excellent. The inferior agreement between the experimental and calculated equiensemble energy differences $\Delta\mathcal{E}_M$, which are more visible in Figure 7.2., accounts for the deviations ($\approx 20\%$) between the experimental excited state energies and those calculated by either method.

To emphasize that these deviations are due to the non-uniform dependence of the QLDA on M, the figure displays on the right vertical axis the excited-state energies averaged over all angular momenta associated with each principal quantum number n. This averaging eliminates the fluctuations in the accuracy of the QLDA and produces excellent agreement with the averaged experimental energies.

Table III: Eleven lowest energies[a] in the spectrum of the He atom, calculated by the equiensemble (EQ), fractional occupation (FO), and transition state (TS) methods, compared with the experimental values[b] (EXPT).

Multiplet	I	ΔE_I^{EXPT}	ΔE_I^{EQ}	ΔE_I^{FO}	ΔE_I^{TS}
1S	1	1.807	1.947(8%)	1.972(9%)	
2S	2	0.336	0.422(26%)	0.422(26%)	0.263(22%)
2P	3	0.262	0.244(7%)	0.243(7%)	0.171(35%)
3S	4	0.134	0.171(27%)	0.174(29%)	0.096(28%)
3P	5	0.115	0.123(7%)	0.118(3%)	0.076(34%)
3D	6	0.111	0.093(16%)	0.086(23%)	0.067(40%)
4S	7	0.072	0.086(19%)	0.079(10%)	0.051(29%)
4P	8	0.064	0.075(17%)	0.073(14%)	0.040(38%)
4D	9	0.063	0.059(6%)	0.060(5%)	0.037(41%)
4F	10	0.063	0.048(24%)	0.047(25%)	0.034(46%)
5S	11	0.045	0.047(4%)	0.045[c]	0.032(30%)

[a] All energies (Ry) subtracted from the ionization threshold, *i.e.*, $\Delta E_I \equiv \overline{E}_I - E_1^+$.

[b] Percentages show deviations from experimental energies.

[c] Reference energy for fractional occupation approach.

Table III and Figure 7.2. also compare the two density-functional approaches with Slater's transition-state method. By substituting the ground-state energy functional for the ensemble energy functional, the latter avoids the numerical problem just discussed and produces more uniform accuracy. In spite of this, as the table and the figure indicate, the equiensemble and fractional occupation approaches are more reliable.

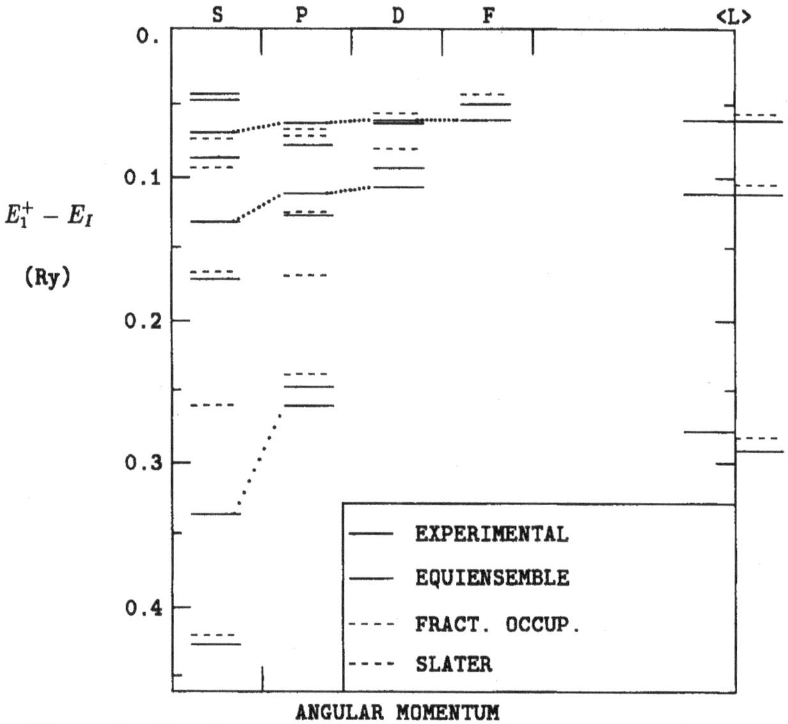

Figure 1: Ten lowest excitation energies for He atom, calculated by the two density-functional approaches described in the text, compared with the experimental values and the results of Slater's transition state method.

8. CONCLUSIONS

The formally exact expressions in Section 5., complemented by the approximations in Section 6., define two ensemble approaches to the calculation of excitation energies. Both schemes, the equiensemble and the fractional occupation procedures, have been applied to the He atom (Section 7.).

For large ensemble multiplicity, the QLDA is very reliable. Its accuracy is, nevertheless, M-dependent. As a consequence, in the equiensemble approach the equiensemble-energy differences contain large percent errors; through Eq. (55) these are subsequently transmitted to the calculated excited state energies.

In principle, the fractional occupation procedure can avoid this problem. To be practical, however, that approach depends on approximations for the derivative $\partial E_{xc}^{W}[\rho^{W}(r)]/\partial w$. The illustrative form proposed in Section 7.2. involves two approximations: (1) neglecting the terms of $\mathcal{O}(1/M^2)$ in Eqs. (66) and (67) and (2) substituting the QLDA form (64) for $E_{xc}^{W}[\rho^{W}]$. As the numerical results indicate, the former approximation is excellent. The latter, however, suffers from the limitation of the QLDA, which makes the fractional occupation and equiensemble approaches essentially equivalent.

In summary, the two procedures produce fairly accurate excited state energies (typical error of 20%). Both methods admit improvements, provided only that (1) a uniformly accurate approximation, to replace the QLDA, or (2) a better approximation for $\partial E_{xc}^{W}[\rho^{W}]/\partial w$ be developed. Unfortunately, the uniformity in the former alternative imposes a very unwieldy constraint. The second research line seems more promising.

ACKNOWLEDGMENTS

Hardy Gross and Walter Kohn contributed many of the ideas in this paper. Most others were shaped in afternoon discussions we had in Walter's ocean-view office in Broida Hall. Collaborating with them was an honor as well as a great pleasure; I am deeply indebted to their incentive and hospitality. The NSF (USA, contract DMR-87-03434), the CNPq, and the FAPESP (Brazil) supported this work.

REFERENCES

1. P. Hohenberg and W. Kohn, Phys. Rev. 136, B864-871 (1964).
2. A. K. Theophilou, J. Phys. C 12, 5419 (1978); N. Hadjisavvas and A. K. Theophilou, Phys. Rev. A 32, 720 (1985).
3. W. Kohn, Phys. Rev. A 34, 737 (1986).
4. E. K. U. Gross, L. N. Oliveira, and W. Kohn, Phys. Rev. A 37, 2809 (1988).
5. For a review, see W. Kohn and P. Vashishta, in *Theory of the Inhomogeneous Electron Gas*, edited by S. Lundqvist and N. H. March (Plenum, New York, 1983), p. 79.
6. A. C. Pedroza, in *Proceedings of the First Brazilian School on Electronic Structure* (Universidade de Brasília, 1989).
7. L. J. Sham and W. Kohn, Phys. Rev. 145, 561 (1966).
8. O. Gunnarsson and B. I. Lundqvist, Phys. Rev. B 13, 4274 (1976).
9. H. Englisch and R. Englisch, Physica A 121, 253 (1983).
10. N. D. Mermin, Phys. Rev. 137A, 1441 (1965).
11. M. W. C. Dharma-wardana and R. Taylor, J. Phys. C 14, 629 (1981); F. Perrot and M. W. C. Dharma-wardana, Phys. Rev. A 30, 2619 (1984); S. Tanaka, S. Mitake, and S. Ichimaru, Phys. Rev. A 32, 1896 (1985); R. G. Dandrea, N. W. Ashcroft, and A. E. Carlsson, Phys. Rev. B 34, 2097 (1986).
12. R. Courant and D. Hilbert, *Methods of Mathematical Physics* (Interscience, New York, 1965), Vol. I, p. 459.

13. E. K. U. Gross, L. N. Oliveira, and W. Kohn, Phys. Rev. A 37, 2805 (1988).

14. J. Katriel, J. Phys. C 13, L375 (1980), has identified the equiensemble with the M-particle ground-state of a "super-Hamiltonian". With this identification, the ensemble of fractionally occupied states corresponds to an ensemble of open systems [see J. P. Perdew, Int. J. Quantum Chem. Symp. 19, 497 (1986) and references cited].

15. J. C. Slater, *The Self-Consistent Field for Molecules and Solids* (McGraw-Hill, New York, 1974), Vol. IV, p. 51.

16. M. Levy, Phys. Rev. A 26, 1200 (1982).

17. E. H. Lieb, in *Physics as Natural Philosophy*, edited by A. Shimony and H. Feshbach (MIT, Cambridge, 1982), p. 111.

18. W. Kohn and L. J. Sham, Phys. Rev. A 140 1133 (1965).

19. D. C. Langreth and J. P. Perdew, Phys. Rev. B 15, 2884 (1977).

20. see, *e.g.*, A. L. Fetter and J. D. Walecka, *Quantum Theory of Many-Particle Systems* (McGraw-Hill, New York, 1971), p. 28.

21. S. H. Vosko, L. Wilk, and M. Nusair, Can. J. Phys. 58, 1200 (1980).

22. D. M. Ceperley and B. J. Alder, Phys. Rev. Lett. 45, 566 (1980).

23. B. Y. Tong and L. J. Sham, Phys. Rev. 144, 1 (1966).

24. J. P. Perdew and A. Zunger, Phys. Rev. B 23, 5048 (1981).

25. To avoid the cumbersome notation $\rho'^{W}(r)$, the ensemble density associated with the potential $v'(r)$ is denoted $\rho'(r)$.

26. L. N. Oliveira, E. K. U. Gross, and W. Kohn, Phys. Rev. A 37, 2821 (1988).

27. W. C. Martin, J. Phys. Chem. Ref. Data, 2, 257 (1973).

IMPURITY STATES IN LOW AND THREE DIMENSIONAL SYSTEMS

A. Ferreira da Silva

Instituto de Física - Universidade Federal da Bahia
Campus da Federação
40 210 Salvador, Bahia, Brazil

ABSTRACT

We present a brief review of our current understanding on the impurity states in low and three dimensional semiconductor systems. The metal-insulator transition, electric conduction, specific heat, spin susceptibility, optical absorption, refractive index and cathodoluminescence are investigated in the light of experimental findings.

1. INTRODUCTION

Many of the studies to date have focussed on the extrinsic properties of the low and three dimensional material systems. Specifically, the impurities have shown to be very significant to probe the nature of the metal-insulator transition in disordered systems as well as in determining the optical and electronic properties of doped semiconductors and artificially structured solids [1-7]. The later ones belong to a new class of man-made solids composed of thin layers of different materials, i.e., semiconductor microstructures with extremely, small spatial extensions [5-8].

In this brief review we will attempt to give a phenomenological description, dimensional analysis and clues from our current understanding of the interactions between shallow impurities in semiconductor structures.

2. IMPURITY STATES IN SEMICONDUCTOR SYSTEMS

In solid state physics we are familiar with the properties of crystalline structures, in which the one-electron wave functions are the Bloch functions extended throughout the whole structure. The study of the electronic properties of such structures is simplified by the periodicity of the lattice which enables us to reduce the problem to the solution of the Schrödinger equation in a single unit cell. In Fig. 1 we show schematically the different density of states D (E) associated with the electrons confined to a semiconductor layer or slab, e.g., low (1 D and 2 D) and three dimensional (3 D) systems.

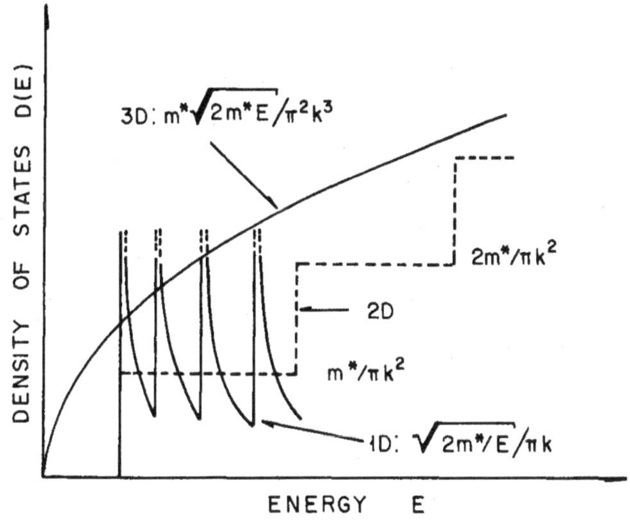

Fig. 1. Schematic view of the density of states for 1 D, 2 D and 3 D systems.

The continuous 3 D D(E) per unit of volume is related to the square-root energy, while for 2 D the D(E) per unit area is presented as constant discret subbands, which can be split into sharply peaked 1 D structure (considering the carriers confined to moving freely in a direction with the motion quantized in the transverse directions), given rise to a square-root singularity of equal strength[8].

Since electronic processes in semiconductors and semiconductor devices usually only involve charge carriers associated with states close to the bottom of the conduction band CB (electrons) or to the top of the valence band (holes) there is less need to have a complete knowledge of the band structure and corresponding Bloch functions. Then the model we use to investigate the impurity states is the usual one, where we regard the extra charge on a donor impurity atom as forming a Coulomb centre on which the electron is bound with a hydrogenlike wave function. As an example we consider bulk silicon (Si) for which the energy gap is 1.2 e V, the dielectric constant is $\varepsilon_{Si} = 11.4$ and the average effective mass is about 0.2 of the free electron value. Doping with phosphorus (P) and/or arsenic (As) produce/s shallow impurity levels of 0.045 eV for P and 0.053 e V for As below the conduction band, and so we picture the electrons in hydrogenlike 1s state with an effective Bohr radius of 15 Å and 14 Å for P and As respectively in Si. The results are the doped systems Si : P, Si : As or Si : P,As. Increasing doping the metallic state occurs at densities above the critical concentration for metal-insulator transition, i. e. , $N_c^P = 3.7 \times 10^{18}$ cm^{-3} and $N_c^{As} = 8.5 \times 10^{18}$ cm^{-3} [9]. At these concentrations, the donors would be 65 Å (P) and 50 Å (As) apart if they were distributed over a regular lattice (SC, BCC, FCC and Diamond) of the host material and averaged these different arrangement to the donors [10,11]. Since the interatomic distance for Si is 2.35 Å, we see that the orbit of the electron is sufficiently large that the lattice, which is very little distorted by the electron, justifying then the use of the effective mass theory. In Fig. 2 we show schematically the P and As donors occupying substitutional positions in Si.

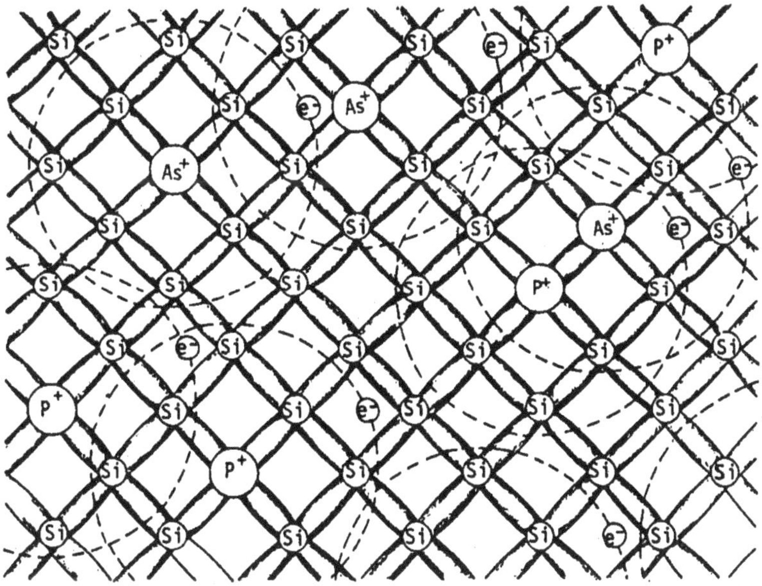

Fig. 2. A schematic view of phosphorus and arsenic donors occupying substutional positions in silicon. (Adapted from K. - F. Berggren, Ref. 3).

3. IMPURITY BAND FORMATION

3.1. 3 D Impurity States

3.1.1. Metal-insulator transition. Using an one-band formalism the impurity density of states can be obtained by the one-electron Green function:

$$(E-H)\ G(\vec{r},\vec{r}\,') = \delta(\vec{r}-\vec{r}\,'), \qquad (1)$$

where H is the one-electron Hamiltonian.

We expand $G(\vec{r},\vec{r}')$ in terms of a set of Kohn-Luttinger (KL) wave functions

$$F_i(\vec{r}) = \frac{1}{\sqrt{\gamma}} \sum_{\beta=1}^{\gamma} \phi(\vec{r}-\vec{R}_i) \exp[i\vec{k}_\beta.(\vec{r}-\vec{R}_i)], \tag{2}$$

where γ is the number of equivalent minima, $\phi(\vec{r})=(\alpha^3/\pi)^{1/2} \exp(\alpha r)$ is the 1s hydrogenic envelop function, and $\alpha^{-1} = a^*$. γ is 4 and 6 respectively for Ge and Si. We then obtain:

$$G(\vec{r},\vec{r}') = \sum_{ij} F_i(\vec{r}) G_{ij} F_j(\vec{r}), \tag{3}$$

where G_{ij} are the unknown coefficients of the expansion. We can easily get

$$\sum_{\ell}(E S_{i\ell} - H_{i\ell})G_{\ell m} = \delta_{im}, \tag{4}$$

where:

$$S_{i\ell} = \int F_i^*(\vec{r})F_\ell(\vec{r})d\vec{r}, \tag{5}$$

and

$$H_{i\ell} = \int F_i^*(\vec{r})|H|F_\ell(\vec{r})d\vec{r} = E_I S_{i\ell}+V_{i\ell}. \tag{6}$$

Here E_I is the ionization energy of an impurity level as measured from the bottom of the CB, $S_{i\ell}$ is the overlap, and $V_{i\ell}$ is the hopping matrix elements.

By applying the well known Matsubara-Toyozawa (MT) diagrammatic method [12] (for more details the reader should refer to both original works of MT), we can express the configuration averaged one-electron density of state D(E) [13] as:

$$D(E) = -\frac{1}{\pi} \mathrm{Im}\left(\overline{G_{ii}} + \frac{N}{8\pi^3} \frac{\zeta}{E'^2} \int \frac{S(\vec{k})T(\vec{k},E')}{1 - \frac{N\zeta}{E'} T(\vec{k},E')}\right)d\vec{k}, \tag{7}$$

where $E' = E - E_I$ and N the impurity concentration.

If one neglects the crossed terms relative to different minima in Eqs. 5 and 6, we can write the terms appearing in Eq. 7 as:

$$\alpha^3 T(\vec{k},E') = -\frac{32\pi}{\gamma}\sum_{\ell}\frac{1}{(1+|\frac{\vec{k}-\vec{k}_\ell}{\alpha}|^2)^3} - E'\alpha^3 S(k) \tag{8}$$

where:

$$\alpha^3 S(\vec{k}) = \frac{64\pi}{\gamma}\sum_{\ell}\frac{1}{(1+|\frac{\vec{k}-\vec{k}_\ell}{\alpha}|^2)^4} \tag{9}$$

The banwidth ΔW is obtained alter some manipulation from Eq. 7. We use also $\Delta W/U = 1.15$ as the criterium for the critical concentration [10,11].

Fig. 3 represents the dependence of band edges (E_ℓ = lower energy and E_u = upper energy) and bandwidth ΔW of the impurity bands of Ge and Si, on $N^{1/3}a^*$. The position of E_I is set at the origin. The crossing of the dot-dashed line (1.15U) with the dotted lines (ΔW) indicates the values of $N^{1/3}a^*$ at the transition, for both Si and Ge. E(N) refers to nonoverlap effects, E(O) refers to overlap effects. ΔW remains almost unaltered for both cases, given approximatively the same value for the transition.

3.1.2. Specific heat. We have also taking into account the valley-splitting in the calculation of D(E) [14]. Furthermore we have used it to get the low-temperature specificheat. The results are shown in Fig. 4 as a function of temperature for concentrations around the critical concentration N_c.

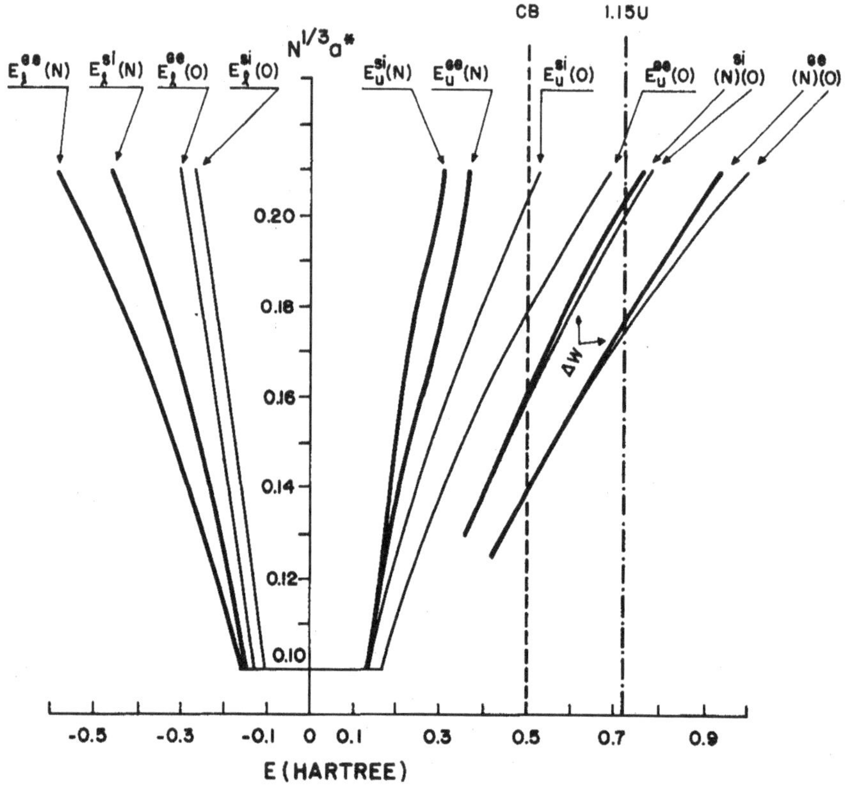

Fig. 3 Dependence of band edges (E_ℓ is the lower energy and E_u the upper energy) on $N^{1/3}a^*$. The position of E_I is set at the origin. The bottom of the conduction band (CB) is set at 0.5 effective Hartree. See text for details.

At very low temperatures, i.e. below 1K, and impurity concentrations below N_c, one observes a characteristic anomaly. Recent cluster models have reproduced fairly well such characteristic features of the experiments in Si : P [16,17].

3.1.3. Impurity resistivity. Assuming the system to be uncompensated the impurity resistivity $\rho = 1/\sigma$ (where σ is the conductivity), at zero temperature, is calculated numerically at the Fermi energy E_F from the

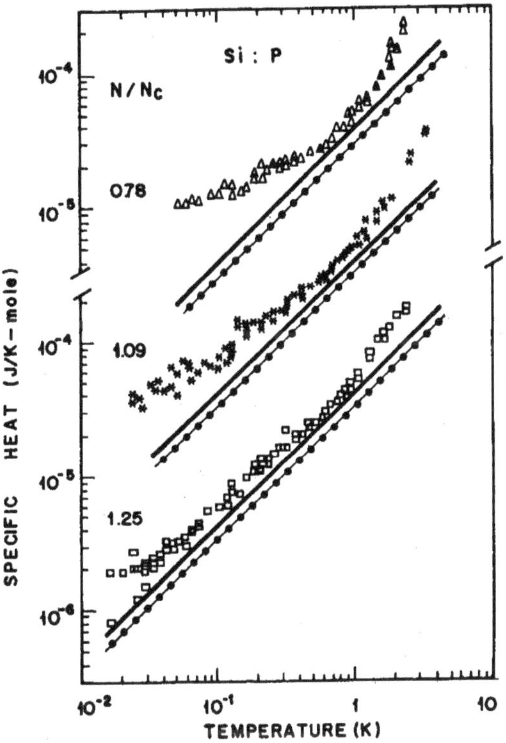

Fig. 4 Specific heat of Si:P as a function of temperature for concentrations around N_c. The data are represented by triangles, stars anq squares [15]. The dotted lines with full circles correspond to the undisturbed silicon conduction band. The solid lines are the theoretical calculation for valley-splitting.

well known Kubo's equation [18]:

$$\frac{1}{\rho^{\gamma}} = \frac{2\pi e^{2}\nu}{\lambda \hbar^{2}} \int \Sigma(E)\ [-df(E)/dE]\, dE\ , \tag{10}$$

where γ = 3D or 2D, ν is the valley degeneracy, λ is the isotropy of the system. Σ (E) is a function of the Fourier transform of $V_{i\ell}$, impurity concentration, averaged diagonal Green function and the pair correlation function [18]. One should note that Eq. 10 is directly, affected by ν [15]. This is due to the fact that when of its formulation one arrives at the dependence of $\nu^2 D^2(E_F)$ and at the average of the matrix elements between eigenfunctions, very close to E_F - which is reduced by ν, thus leading at the end to such enhancement factor. In Fig. 5 we present different results for resistivity of Si:P as function of N, with ν = 6, γ = 3D and λ = 3.

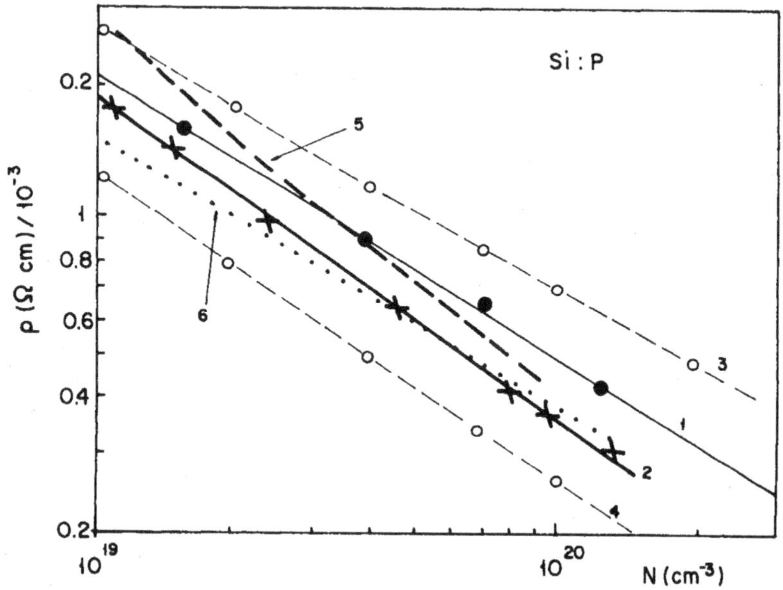

Fig. 5 The impurity resistivity ρ as a function of concentration N. Line 1 with full circles represents experimental data from Ref. 20, and line 2 with crosses are data from Ref. 21. Dashed lines 3 and 4 with open circles represent resistivities calculated with different potentials by an ionized-impurity scattering scheme, Ref. 22. Dashed line 5 represents our calculation with an Alternant Molecular Orbital (AMO) method and Eq. 10, while dotted line 6 is the result of a correlated electrongas system, Ref. 23.

3.1.4 Spin susceptibility. The influence of the many-valley isotropic effect of the CB [11, 24], temperature and correlation are also investigated in the calculation of spin susceptibility χ_s for the systems Si:P, Si:As, Si:Sb and Si:P, As (this later system will be described below). The χ_s obtained [25] - by an extension of Gutzwiller method done by Chao and Berggren[26], is written as

$$\chi_s = \eta_\chi (T)\, \chi_o (T), \tag{11}$$

where η_χ is an enhancement factor and χ_o is the Pauli spin susceptibiliy. The results are shown in Fig. 6.

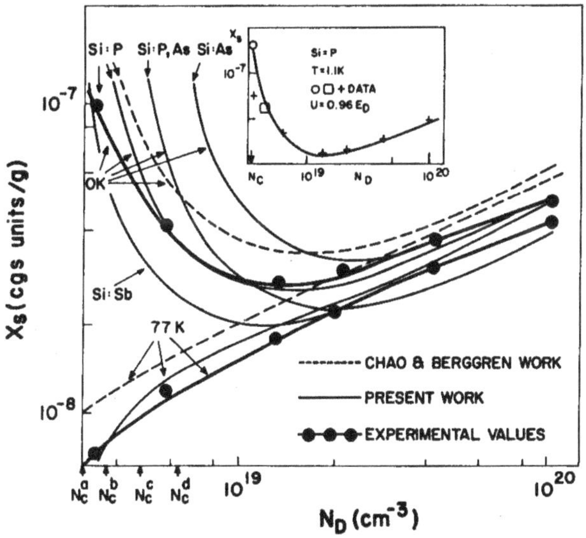

Fig. 6 The spin susceptibility χ_s as a function of concentrations for different systems. At 77 K χ_s is for Si:P. In the inset χ_s is for T = 1.1K, U is the correlation energy scaled to the ionization energy, and the experimental values are from Ref. 6.

3.1.5 Optical absorption. The optical absorption, associated to the impurity density of states, due to interacting donor-pairs are also investigated. It is derived from the transition $H_2 \rightarrow H_2^+$ of their lowest states. In our approximation these are the states in which one of the electrons is promoted from the donor-pairs cluster to the bottom of the CB [27]. For the H_2 - molecule we hawe used the AMO method taking into account the valley degeneracy of Si. The absorption in this pair approximation has the form

$$\alpha\ (E) \simeq \int P_{nn}\ (R)\ \delta\ [E - E_e^o\ (R)]\ dR, \qquad (12)$$

where $P_{nn}(R)$ is the nearest-neighbor Poisson distribution and $E_e^o\ (R)$ is the excitation energy. The theoretical results compared to the experiment [28] are shown in Fig. 7.

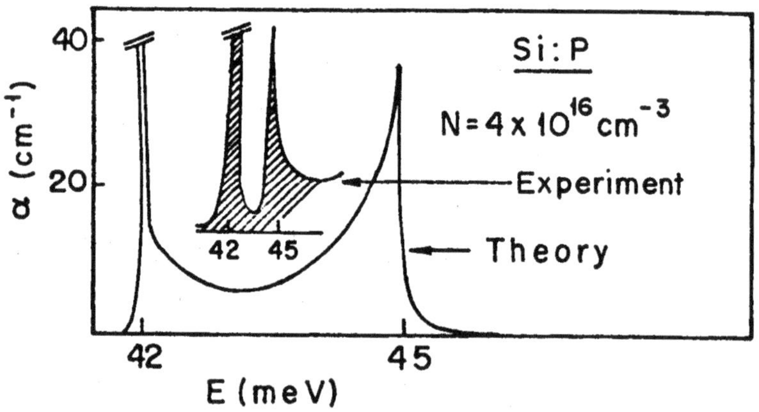

Fig. 7. Optical absorption due to donor-pairs as a function of energy for N = $4\times10^{16}cm^{-3}$ in the Si:P system. Here we take the

The refractive index, of Si:P as dependent on doping concentration is obtained using the Kramers-Kronig relation [29].

$$n(0) = n_{Si} + (c/\pi) \int_0^\infty [\alpha(E)/E^2]\, dE, \tag{13}$$

where $n_{Si} = \epsilon_{Si}^{1/2} = (11.4)^{1/2}$, $n(0)^2 = \epsilon$ and $\alpha(E)$ comes from Eq.12. In Fig. 8 we show the refractive index RI as a function of N for the pair approximation. For very low concentration the system leads to the expected $n_{Si} = (11.4)^{1/2} = 3.38$, while for N around 10^{17}, n(0) starts to increase, rising most rapidly, given an asymptotic behavior for $N \to N_c$. Such results support some previous measurements [30].

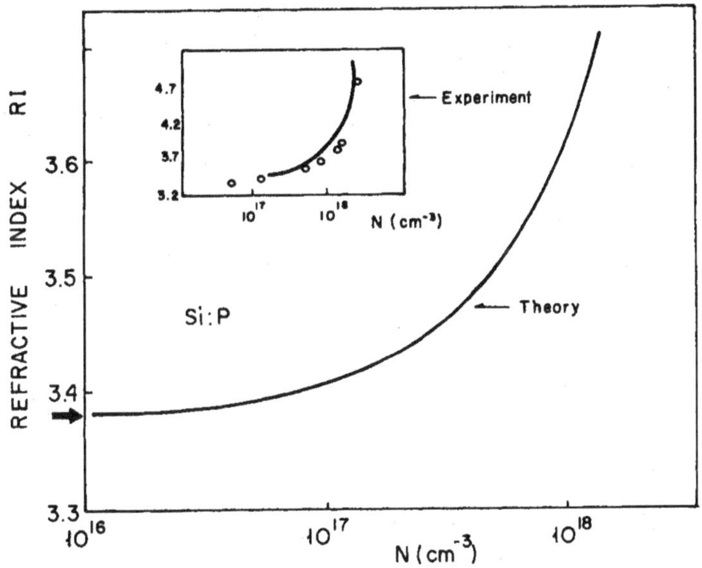

Fig. 8. The refractive index RI of Si:P as a function of impurity concentration N. In the inset, the solid line is from Castner et al., Ref. 31, and open circles are from Townsend, Ref. 30.

3.1.6. Shallow double-donor systems. The work to date has concentrated much effort of investigating single-donor states in doped semiconductors. By now there is a report on metal-insulator transition in a shallow double-donor system, namely Si:P, As, obtained by Newman and Holcomb (NH) [32] in their measurements of electrical conductivity. They found no measureable effects of the additional vertical disorder introduced by the different impurity binding energies, i.e., of Si:P and Si:As, in comparison to these two single-donor systems. The electron-electron interaction effects were kept unchanged. The schematic view of these donors is shown in Fig. 2. A first attempt to confirm such observable results was given recently by a Hubbard's tight-binding scheme and a cluster approach [24]. It was found that the impurity bandwidth, density of states and, the physical behavior of the spatial distribution of the eigenvectors are related to an effective hopping energy, scaled to the ionization energy of Si:P, written as

$$T = \frac{e^2}{\varepsilon_{si} a^*_P} \left(V + \frac{S}{2}\right) . I, \tag{14}$$

where a^*_P is the effective Bohr radius of P in Si.

I is an interference factor due to the valley degeneracy. The expression for S and V are obtained as

$$S = \frac{8y^{5/2}}{(1 - y^2)^2} \quad 1 + \exp\,(1 - y)R^* + \frac{1 - \exp(1 - y)R^*}{(1 - y^2)R^*} \quad \exp\,(- R^*) \tag{15}$$

and

$$V = \frac{4y^{1/2}}{(1-y^2)} \quad 1 + \frac{1 - \exp^{p}(1-y)R^*}{(1-y^2)R^*} \quad \exp(-R^*) \tag{16}$$

where $y - a^*_x/a^*_p$ and $R^* = y - R/a^*_p$, R being the separation between donor states.

It is worth to mention that the different binding energies introduced by a double-donor system give a different hopping energy when compared to a single-donor system, e.g., at a certain concentration the bandwidths are different and certainly will be the shape of the impurity bands for both systems, which leads to different N_c. For the sake of comparison we plot in Fig. 9, for a SC lattice, T as a function of N for the systems Si:P, Sb, Si:P and Si:P, As. We note the difference in T, caused by the different binding energies, in Si:P, Sb and Si:P, As, when compared to Si:P, leading to a transition in the following order $N_c^{P,Sb} < N_c^{P} < N_c^{P,As}$. It was found $N_c^{P,Sb} = 3.4 \times 10^{18}\ cm^{-3}$, $N_c^{P} = 3.9 \times 10^{18}\ cm^{-3}$ and $5.4 \times 10^{18}\ cm^{-3}$ [33]. The experimental data are $N_c^{P} = 3.7 \times 10^{18}\ cm^{-3}$ and $N_c^{P,As} = 5.1 \times 10^{18}\ cm^{-3}$ [32]. There is no data for $N_c^{P,Sb}$.

The density of states for these critical concentrations, which are indicated in Fig. 9, are shown in Fig. 10. We approximate the D(E) by a semicircular form.

3.2. Low Dimensional Impurity Systems

The study of charge carriers in low dimensional (2 D and 1 D) electronic systems has been a subject of considerable scientific and technological interest recently [7,8,34]. An good example for investigations of impurity states in such systems is the MOSFET (Metal - Oxide - Semiconductor Field Effect Transistors) devices. Because of the electrostatic confining potential in the Si at the Si O_2 interface ,

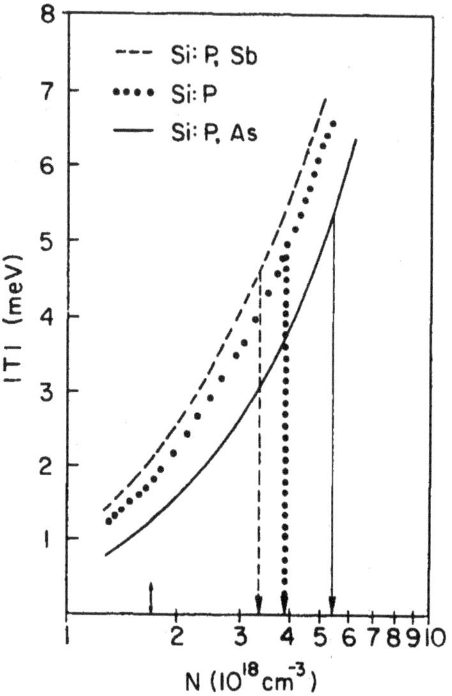

Fig. 9. Hopping energy |T| as a function of concentrations N for different systems. The arrows indicate the N_c for each system.

electron motion perpendicular to the interface is quantized into sets of subbands with the motion parallel to the interface unaffected, thus leading to an essentially two-dimensional conductor. The situation above is the result of the band bending, due to the electric field over the semiconductor - oxide interface. When the bottom of the CB dip bolow E_F (fixed by the host properties), the electrons enter the potential well at the surface forming an inversion layer. At low electron density and low temperature (for the case that E_F only intersects one subband) one has the electric quantum limit. Below this level one has the impurity

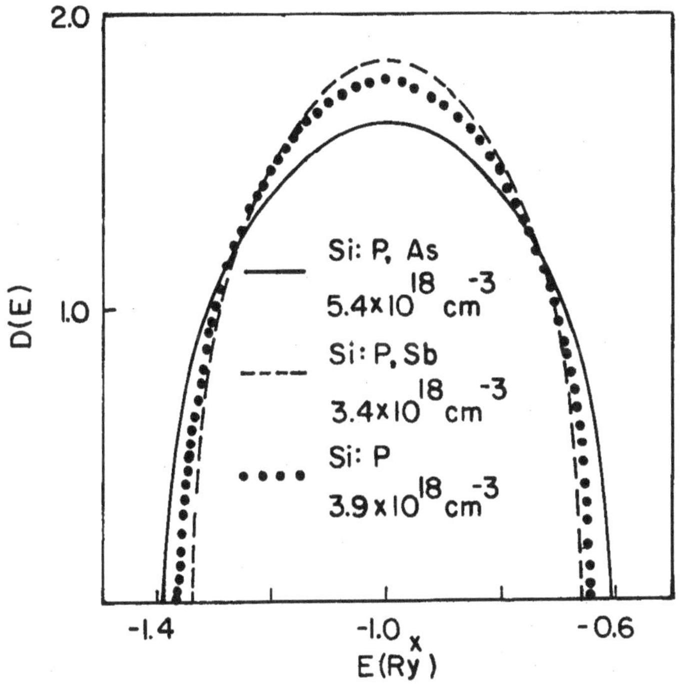

Fig. 10. Density of states for different systems at N_c, with the hopping energy indicated in Fig. 9.

Level. With increasing impurity concentration, this level will broaden and eventually merge at a certain concentration. In Fig. 11 we show schematically such situations, which are also valid for Quantum-Well (QW) and Quantum-Well Wires (QWW).

Donor impurities can be made to occur in Ga As - $Ga_{1-x}Al_xAs$ heterostructure as shown in Fig. 12 [35].

For one donor either inside the small-band gap region (Ga_{1-x} Al_x As), an additional electron state appear below the first subband then an impurity band will form as the increasing of the impurity concentration.

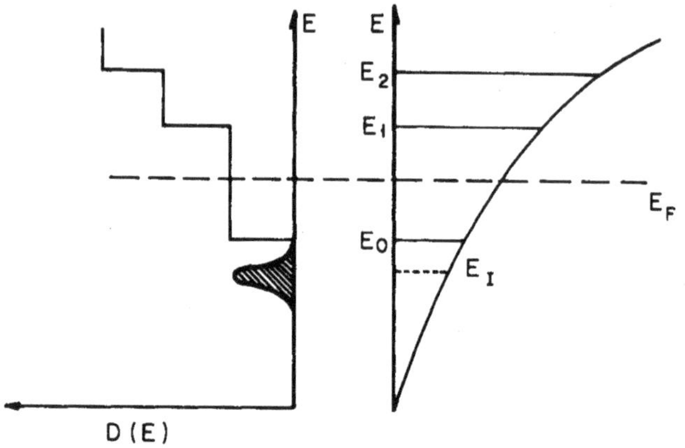

Fig. 11. Schematic view of the quantized energy levels Ei; i=0,1,2, impurity level E_I and the corresponding density of states and impurity density of states in the inversion layer system.

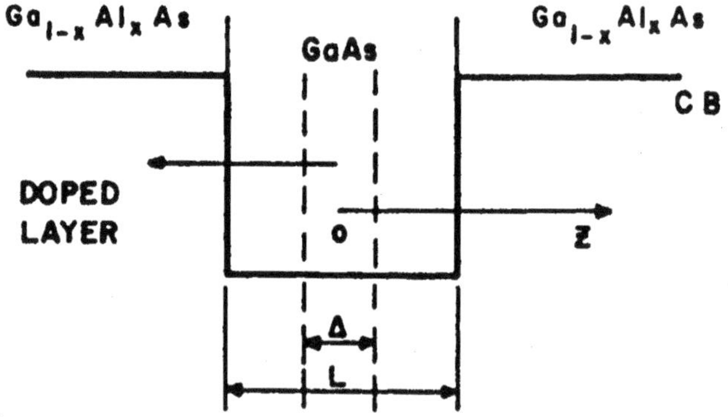

Fig. 12. Scheme of the Ga As - $Ga_{1-x}Al_xAs$ QW. The well width L, the impurity layer thickness, and the z direction are assigned with the origin at the center of the well. Ref. 35

For QWW the situation is similar, i.e., in this heterostructure the electronic motion is quantum mechanically confined on the plane XY while being free to move in the Z direction, along which quasi - one - dimensional (1D) transport can orcur.

In fig. 13 we show, the peak energy $E_P(L)$, obtained at the maximum of D(E), i.e. $D(E_P)$, and the bandwidth $\Delta E(L)$ for $N=10^4\ cm^{-1}$, $E_B(L)$ (binding energy) and $E_P(N)$ for L=200Å. It is worth to say that $E_B \to 5.3meV$ for large value of L (L>>a*). It corresponds to E_B of a donor in GaAs. For $L \to 0$ the electron wave function will recover $Ga_{1-x}Al_xAs$, [36,37]. For the sake of comparison L = 100 Å corresponds to E_B = 24meV, a similar value found by Brum [38]. Lee and Spector [39] found 17meV and Bryant, [40] for a cylindrical cross section, found 16meV. It is also noticed that for L=200Å the system undergoes a similar energy around 13meV for all calculated properties. This bring our attention to the observed cathodoluminescence in the $GaAs\text{-}Ga_{1-x}As_cAl$ heterojunction with L=200Å, by Petroff et al [41]. They observed a broader linewidth of 15meV, with a full width at half-maximum. We have obtained for it 12meV. The peak energies $E_P(L)$ and $E_P(N)$ decrease with a small difference between then. For some approaches which use the center of gravity $E_{CG}(L)$ instead, as function of L, and have impurity bands with many structures [42], one can expect $E_{CG}(N)$ to have the same behavior as for $E_{CG}(L)$.

In previons works for 2D systems, a reasonable agreement between the conductance maximum, found by Ando et al [7], and Hartstein and Fowler [43], and the calculated $D(E_P)$ was obtained [44]. For the above calculation $D(E_P)$ is around $10^5 cm^{-1} meV^{-1}$. No experimental data are available at the moment. Turning to the purely 2D limit, we have calculated the

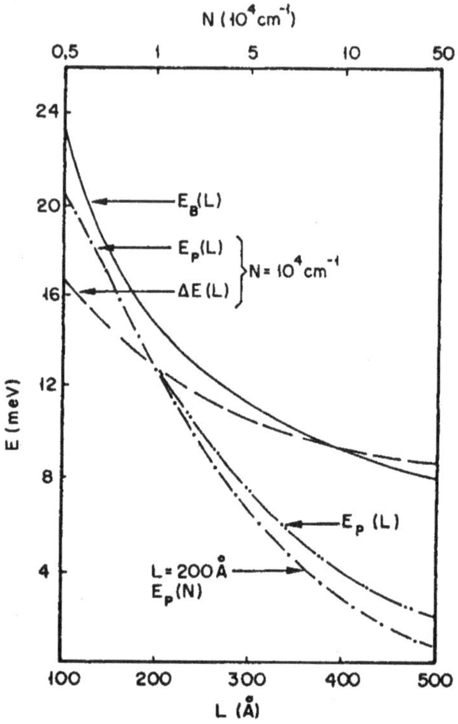

Fig. 13. Variation of binding energy $E_B(L)$, peak energy $E_P(L)$ of maximum D(E) and bandwidth $\Delta E(L)$ with the wire dimension L. Upper scale represents $E_P(N)$ for L=200Å, as a function of impurity concentration N.

impurity resistivity ρ^{2D} for the the sake of comparison with ρ^{3D}. We have used the valley degeneracy $\nu = 2$, as done be Kelly and Falicov [45], in Eq. 10. The results are shown in Fig. 14.

The ρ^{2D} is found to be enhanced by almost six order of magnitude over ρ^{3D} [7]. This rough treatment for ρ^{2D} with the impurity located it self at the inversion layer generates 2D hydrogenlike bound states with binding energy equal to $4R_y^*$. For the case of Si:SiO$_2$, $1R_y^* = 42$ meV ,

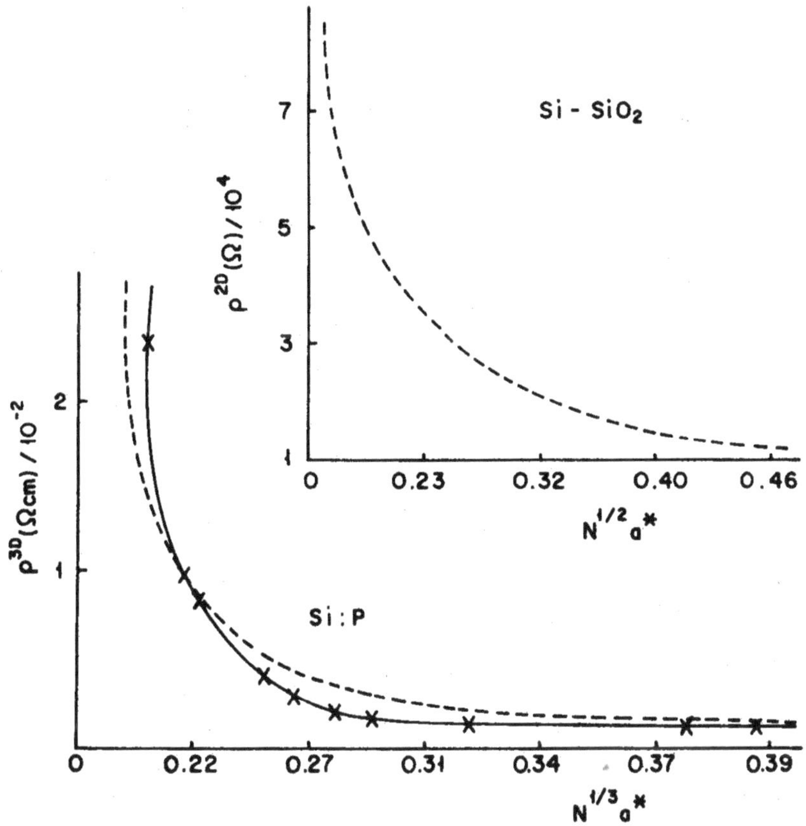

Fig. 14 Impurity resistivities for 3D and 2D systems as a function of $N^{1/3}a^*$ and $N^{1/2}a^*$ respectively. The crosses are the data from Ref. 21. The dashed lines are the present theory.

the effective Bohr radius $a^* = 22\mathring{A}$ and the dielectric constant $\bar{\varepsilon} = (\bar{\varepsilon}_{Ox} + \varepsilon_{si})/2$. For 3D the units were described along the text. Briefly the relationship between the three and low dimensional systems was discussed in the light of various measurements.

REFERENCES

1. "The Metal-Insulator Transition in Disordered Systems", edited by L.R. Friedman and D.P. Tunstall (SUSSP, Edinburgh, 1978).

2. "Current Research on Semiconductor Physics", Proc. of the Second School on Semiconductor Physics, edited by J.A. Leite and C.E.T. Gonçalves da Silva (USP, São Paulo, Brazil, 1985).

3. Ferreira da Silva, A., in Ref. 2, p. 359.

4. Berggren, K.-F., in Ref. 2, p. 387.

5. Stern, F., in Ref. 2, p. 473.

6. "Localisation and Interaction in Disordered Metals and Doped Semiconductors", edited by D.M. Finlayson (SUSSP, Edinburgh, 1986).

7. Ando, T., Fowler, A.B. and Stern, F., Rev. Mod. Phys. 54, 437 (1982).

8. Berggren, K.-F., Int. J. Quantum Chem. 23, 217 (1988).

9. Newman, P.F. and Holcomb, D.F., Phys. Rev. B 28, 638 (1983).

10. Berggren, K.-F., Philos. Mag. 27, 1027 (1973).

11. Ferreira da Silva, A., Phys. Rev. B 37, 4799 (1988).

12. Matsubara, T. and Toyozawa, Y., Prog. Theor. Phys. 26, 739 (1961).

13. Perondi, L.F., Fabbri, M. and Ferreira da Silva, A., Philos. Mag. B 54, 81 (1986).

14. Perondi, L.F., Fabbri, M. and Ferreira da Silva, A., Solid State Commun. (in press).

15. Paalanen, M.A., Graebner, J.E., Bhatt, R.N. and Sachdev, S., Phys. Rev. Lett. 61, 597 (1988).

16. Franzèn, N.-I. and Berggren, K.-F., Philos. Mag. 43, 29 (1981) ; ibid. Phys. Rev. B 25, 1993 (1982).

17. Eto, M. and Kamimura, H., Phys. Rev. Lett. 24, 2790 (1988).

18. Ferreira da Silva, A., J. Phys. C 13, L 427 (1980).

19. Saso, T. and Kasuya, T., J. Phys. Soc. Japan, 48, 1566 (1980).

20. Chapman, P.W., Tufte, D.N., Zook, J.D. and Long, D., Appl. Phys. 34, 3291 (1963).

21. Yamanouchi, C., Mizuguchi, K. and Sasaki, W., J. Phys. Soc. Japan 22, 859 (1967).

22. Csavinsky, P., Beauperthuy, L. and Morrow, R.A., Phys. Rev. 24 , 3602 (1981).

23. Sernelius, B.E. and Berggren, K.-F., Philos. Mag. B 43, 115 (1981).

24. Ferreira da Silva, A., Phys. Rev. Lett. 59, 1263 (1987).

25. Ferreira da Silva, A., Phys. Rev. B 38, 10055 (1988).

26. Chao, K.A. and Berggren, K.-F., Phys. Rev. Lett. 34, 880 (1975).

27. de Brito Mota, F. and Ferreira da Silva, A., Solid State Commun. (in press).

28. Capizzi, M., Thomas, G.A., De Rosa, F., Bhatt, R.N. and Rice, T.M., Phys. Rev. Lett. 44, 1019 (1980).

29. de Brito Mota, F. and Ferreira da Silva, A., Internal Report (to be submitted).

30. Townsend, P., J. Phys. C 11, 1481 (1978).

31. Castner, T.G., Lee, N.K., Cieloszyk, G.S. and Salinger, G.L., Phys. Rev. Lett. 34, 1627 (1975).

32. Newman, P.F. and Holcomb, D.F., Phys. Rev. Lett. 51, 2144 (1983).

33. Ferreira da Silva, A., Phys. Scr. T 14, 27 (1986).

34. Berggren, K. -F., Thornton, T.J., Newson, D.J. and Pepper, M., Phys. Rev. Lett. 57, 1769 (1986).

35. de Andrade e Silva, E.A., da Cunha Lima, I.C., Ferreira da Silva , A. and Leal, C.E., Phys. Rev. B 37, 8228 (1988).

36. de Andrade e Silva, E.A., da Cunha Lima, I.C. and Ferreira da Silva, A., Phys. Rev. 37, 8537 (1988).

37. Ferreira da Silva, A., Phys. Rev. B (in press).

38. Brum, J.A., Solid State Commun. 54, 179 (1985).

39. Lee, J. and Spector, H.N., J. Vac. Sci. Technol. B 2, 16 (1984).

40. Bryant, G.W., Phys. Rev. B 29, 6632 (1984).

41. Petroff, P.M., Gossard, A.C., Logan, R.A. and Wiegmann, W., Appl. Phys. Lett. 41, 635 (1982).

42. Oliveira, L.E. and Folicov, L.M., Phys. Rev. B 34, 8676 (1986) ; Weber, G. Schulz, P.A. and Oliveira, L.E., Phys. Rev. B 38, 2179 (1988).

43. Hartstein, A. and Fowler, A.B., Phys. Rev. Lett. 34, 1435 (1975); ibid, Bull. Am. Phys. Soc. 28, 322 (1983).

44. da Cunha Lima, I.C., Ferreira da Silva, A. and Fabbri, M., Surf. Sci. 134 (1983); de Andrade e Silva, E.A. and da Cunha Lima, I.C., Phys. Rev. Lett. 58, 925 (1987).

45. Kelly, M.J. and Falicov, L.M., Phys. Rev. Lett. 37, 1021 (1976).

PHONONS IN SEMICONDUCTOR SUPERLATTICES

Paulo V. Santos
Instituto de Física - Universidade Estadual de Campinas - UNICAMP
P.O. Box 6165, 13081 Campinas - SP
BRAZIL

ABSTRACT

A brief review of the main concepts related to vibrational properties of semiconductor superlattices is given in this paper. Using a simple "Kronig-Penney"-like models, it will be shown that in these structures the phonons have properties similar to those of the electronic excitations. In particular, depending on the phonon energy, the phonons can be either confined within individual layers or extended over the whole structure. Different methods are then presented for the calculation of the phonon dispersion in periodic structures, with emphasis on those applicable to acoustic phonons.

1. INTRODUCTION

Semiconductor superlattice structures have been intensively studied in the last years. These structures consist normally of two thin layers of different materials with thicknesses d_1 and d_2 that are repeated periodically on the surface of a substrate, as is indicate in the upper part of Fig. 1. As a consequence, a periodic potential having the same period d as the layer repetition is imposed both for electrons in the conduction band (CB) and for holes in the valence band. Most of the interesting electronic properties of the superlattices stems from the resulting modulation of the conduction and valence band edges, leading to the formation of an artificial lattice. As an example, Fig. 1b shows the band structure of a superlattice composed of two materials 1 and 2 with different band gaps $E_g^{(1)} < E_g^{(2)}$. This is the typical situation of III-V superlattices such as GaAs(material 1)/AlAs(material 2), GaAs/GaAlAs and GaSb/AlSb structures. The lowest energy eigenstates for electrons and for holes resemble those of a particle confined in an one-dimensional potentials well of the lower gap material (material 1): the electrons and holes are strongly confined in the layers 1 and their energy levels are discrete and shifted relative to the band edges of material 1 in bulk form. As the electron (or hole) energy increases, the particles can tunnel through the barriers and the become more and more delocalized. The energy levels broad into bands that are indicated by the dashed regions in Fig. 1b (shown only for electrons). The electronic bands related to the artificial superlattice potential differs considerably from those of the bulk constituent materials even for particle energies exceeding the barrier, and depend critically of the thicknesses of the individual layers.

The purpose of this paper is to review basic concepts concerning phonons in superlattice structures, and, in special, those related to acoustic phonons. For the reader interested in a deeper insight into the properties of optical phonons in superlattices we recommend references [1,2,3,4,5] and, in special, the recent review articles from *Cardona*[6] and from *Jusserand et al.* [7] . We shall see in this paper that phonons in superlattice structures behave in a similar way as electrons and holes. The similarity arises from the wave-like character of both vibrational and electronic excitation. The presentation is organized as follows. In the next section (Sec. 2) a simple linear chain model for the phonon dispersion in superlattices is introduced. This model is used to illustrate the concepts of confined and folded phonons. Section 3 is devoted to acoustic phonons. Here, a continuous elastic model is used to describe acoustic phonon propagation in arbitrary directions in a superlattice. In Sec. 4 a method for the calculation of the phonon transmission through

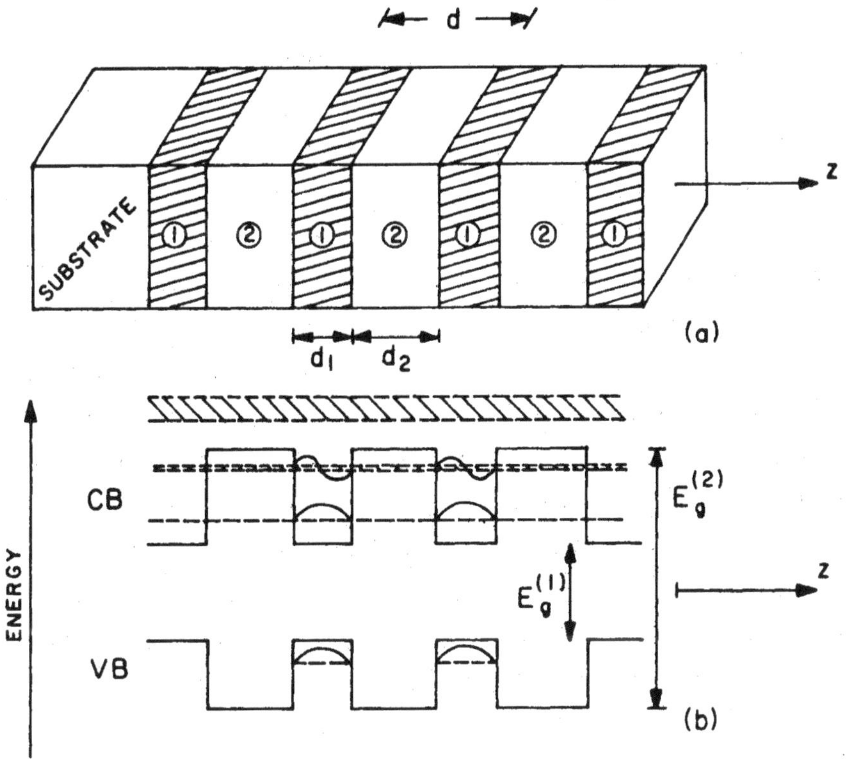

Figure 1: a) Superlattice composed of layers of materials 1 and 2 with thicknesses d_1 and d_2 and, energy gaps $E_g^{(1)}$ and $E_g^{(2)}$, respectively. b) Conduction (CB) and valence band (VB) profile for the the superlattice. The dashed lines limit the superlattice bands and the solid lines within material 1 shows the wavefucntion for strongly localized levels for electrons in the CB and hole in the VB.

multilayer structures is presented and Sec. 5 summarizes the main conclusions of this work.

2. PHONON DISPERSION IN SUPERLATTICES

In order to describe the vibrational properties of superlattices we shall consider the simple model for the atomic configuration displayed in Fig. 2a. The layer materials have a simple cubic lattice with the same lattice constant a and atomic masses m_1 and m_2 for layers 1 and 2, respectively. The layers are stacked in the [001] direction and each period is composed of n_1 atomic planes of material 1 and n_2 atomic planes of material 2. We shall further assume that the vibrational properties of the structure can be describe by taking into account only nearest neighbors interactions (ball and stick model) with force constants that have the same value for both materials. For each bond there are two force constants: a stretching force constant C_l and a bending force constant C_t. Although the last assumptions seem to be very restrictive, the model we are going to develop can explain qualitatively most of the experimental results on the vibrational properties of semiconductor superlattices.

We will consider in the following only vibrational modes propagating in the direction perpendicular to the superlattice layers (z-direction). These are the phonon modes most easily probed experimentally by Raman spectroscopy.[1,9] From the symmetry of the structure it is easy to see that for this propagation direction there are three eigenmodes for each vibrational frequency: a longitudinal mode, where the planes oscillate in the z-direction, and two degenerate transverse modes, where the planes move in the x- and in the y-directions. For these eigenmodes each atomic plane indicated by the dashed lines in Fig. 2a moves as a whole: the forces acting on a single atom located in a plane originates only from its nearest neighbors in nearby planes. The vibrational eigenmodes of the superlattice are therefore the same as those of linear chain shown in Fig. 2b. The force constant C here equals C_l and to C_t for longitudinal and transverse vibrations, respectively.

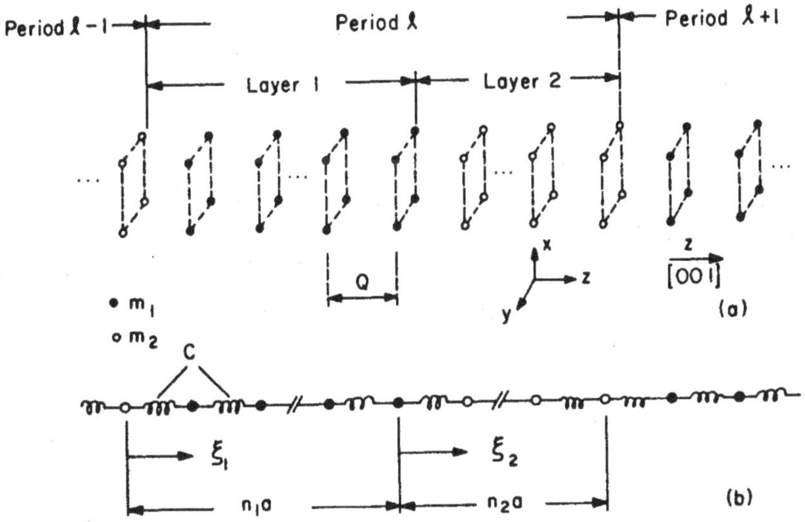

Figure 2: a) Atomic configuration of a superlattice period composed of n_1 monolayers of material 1 (atomic mass m_1) and n_2 layers of material 2 (atomic mass m_2). Both materials have a simple cubic structure with the same lattice constant a. b) Linear chain model obtained by projecting the atomic planes on the z-axis. A nearest neighbor interaction with the same force constant C is assumed for both material.

In order to calculate the vibration eigenmodes of the linear chain in Fig. 2b we will denote by $u_{j,n}^{(l)}$ the displacement from the equilibrium position of the nth atom of the jth layer ($j = 1,2$) of the lth period. $u_{j,n}^{(l)}$ is along the z-direction for longitudinal modes and along the x or y-directions for transverse modes.

Let us consider initially the displacement $u_{j,n}^{(l)}$ of an atom located *within* the jth layer, i.e., $n \neq 1, n_j$. The equation of movement for this atom is:

$$m_j \frac{d^2 u_{j,n}^{(l)}}{dt^2} = C[u_{j,n+1}^{(l)} - u_{j,n}^{(l)}] - C[u_{j,n}^{(l)} - u_{j,n-1}^{(l)}] = C[u_{j,n+1}^{(l)} + u_{j,n-1}^{(l)} - 2u_{j,n}^{(l)}] \quad (1)$$

In the most general case the solution of this equation for each angular vibrational frequency ω is the superposition of two plane waves with amplitudes $a_j^{(l)+}$ and $a_j^{(l)-}$ travelling in opposite direction along the z-axis:

$$u_{j,l}^{(n)}(\omega) = e^{-i\omega t}\left[a_j^{(l)+} e^{ik_j\xi_j} + a_j^{(l)-} e^{-ik_j\xi_j}\right] = e^{-i\omega t} \begin{pmatrix} e^{ik_j\xi_j} \\ e^{-ik_j\xi_j} \end{pmatrix} \underbrace{\begin{pmatrix} a_j^{(l)+} \\ a_j^{(l)-} \end{pmatrix}}_{\mathbf{A}_j^{(l)}} \quad (2a)$$

where

$$\xi_j = n_j a \quad (2b)$$

gives the position of the nth atom in the jth layer. In order to simplify the notation, the coefficients $a_j^{(l)+}$ and $a_j^{(l)-}$ were joined in the two-dimensional vector $\mathbf{A}_j^{(l)}$. The displacement $u_{j,n}^{(l)}(\omega)$ is along the z-direction for longitudinal vibrations and along the x or y-direction for transverse vibrations.

By substituting Eq. (2a) into Eq. (1) we obtain the following relationship between the wave vector k_j and the vibration frequency ω:

$$\sin\frac{k_j a}{2} = \frac{\omega}{\omega_{max}^{(j)}} \qquad \text{where} \qquad \omega_{max}^{(j)} = \sqrt{\frac{4C}{m_j}} \quad (3)$$

Equation 3 constitutes the *dispersion relation* relating the vibration frequency ω and the wave vector k_j of an infinity chain of atoms type j. This relationship is plotted for two materials (1) and (2) with different vibrational properties by the dot-dashed lines in Fig. 3a to 3d. The k_j values are restricted to the interval $-\frac{\pi}{a} < k_j \leq \frac{\pi}{a}$, which defines the first *Brillouin zone* of the one-dimensional chain. The vibration frequencies are confined to values below $\omega^{(j)}_{max}$. Above this frequency the absolute value of the sinus at the left side of Eq. 3 is greater then one and k_j acquires an imaginary component. This would correspond to waves that attenuate strongly while propagating in the chain. Such modes cannot exist in an infinite chain.

Let us consider now the displacements of the two atoms placed *at the interface* between two consecutive superlattice layers j and $j+1$ (see Fig. 2). The atoms have displacements $u^{()}_{j,n_j}$ and $u^{()}_{j+1,1}$, respectively. In the following we drop the period index l since the treatment will be valid both for the interface within a period and between periods. The equation of motion for these atoms will be:

$$\begin{aligned} m_j \frac{d^2 u^{()}_{j,n_j}}{dt^2} &= C[u^{(l)}_{j+1,1} + u^{(l)}_{j,n_j-1} - 2u^{()}_{j,n_j}] \\ m_{j+1} \frac{d^2 u^{()}_{j+1,1}}{dt^2} &= C[u^{(l)}_{j+1,2} + u^{()}_{j,n_j} - 2u^{()}_{j+1,1}] \end{aligned} \tag{4}$$

We now make the assumption that the displacement of the interface atoms is also given by Eq. 2a. Substituting this equation in Eq. 4 we obtain the following relationship relating the displacement vectors $\mathbf{A}^{()}_j$ and $\mathbf{A}_{j+1}l^{()}$ in adjacent layers:[8]

$$\mathbf{A}^{()}_{j+1} = \underbrace{\frac{1}{s_{j+1} - s^{-1}_{j+1}} \begin{pmatrix} f_j(s_{j+1} - s^{-1}_{j+1}) & f_j^{-1}(s_j^{-1} - s^{-1}_{j+1}) \\ f_1(-s_j + s_{j+1}) & f_j^{-1}(-s_j^{-1} + s_{j+1}) \end{pmatrix}}_{\mathbf{M}_j} \mathbf{A}^{(l)}_j \tag{5a}$$

where

$$s_j = e^{ik_j a_j} \quad \text{and} \quad f_j = e^{ik_j d_j}. \tag{5b}$$

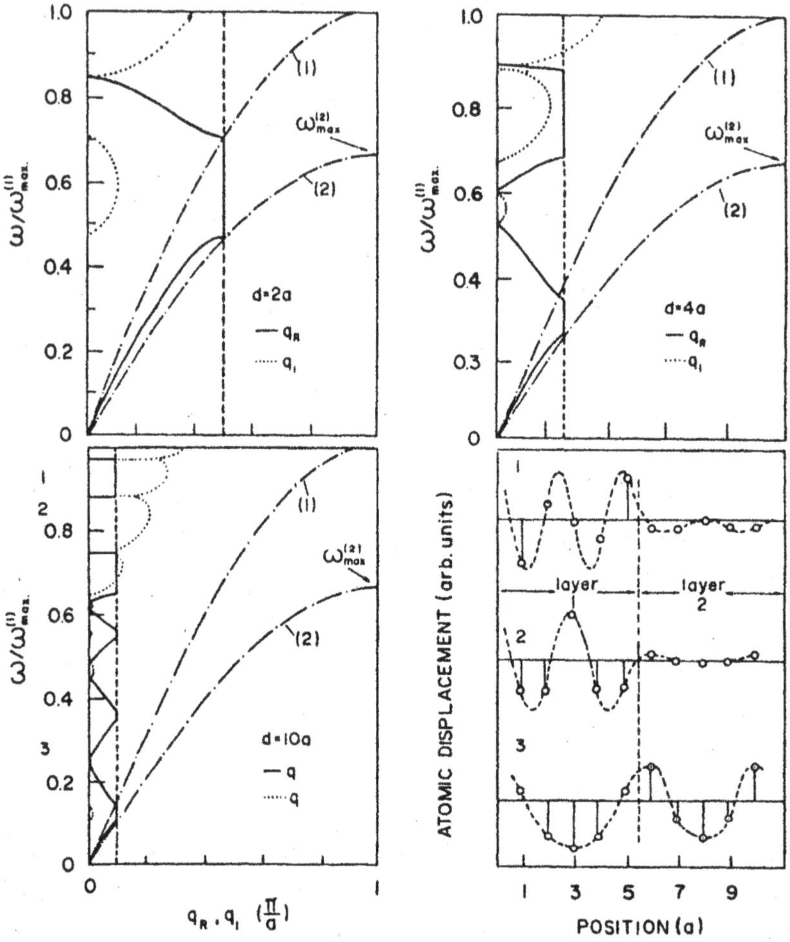

Figure 3: Dispersion relation for phonons propagating along the [100]-direction calculated from the simple linear chain model discussded in the text. The curves are for superlattices with (a) one ($d = 2a$), (b) two ($d = 4a$), and (c) five ($d = 10a$) atoms in each layer. Curve (d) shows the atomic displacement for the zone-center modes indicated by 1, 2, and 3 in curve (c). In each case, the real (q_R) and the imaginary part (q_I) of the phonon wave vector is indicated by solid and dotted lines, respectively. The dot-dashed lines are the bulk dispersions of the layer materials in bulk form.

We return now to the case of the superlattice shown in Fig. 2a. Applying Eq. 5a twice we obtain the following equation relating the atomic displacements in the first layer of successive periods:

$$\mathbf{A}_1^{(l+1)} = \underbrace{\mathbf{M}_2\mathbf{M}_1}_{\mathbf{M}}\mathbf{A}_1^{(l)} \tag{6a}$$

where the elements of the $2x2$ matrix $\mathbf{M}$ are given by:

$$\begin{aligned} M_{11} = M_{22}^* &= \frac{2f_1\{f_2[1-\cos(k_1+k_2)a] - f_2^{-1}[1-\cos(k_1-k_2)a]\}}{4\sin k_1d_1 \sin k_2d_2} \\ &\text{and} \\ M_{12} = M_{21}^* &= \frac{2if_1^{-1}(s_2^{-1}-s_1^{-1})(s_2-s_1^{-1})\sin k_2d_2}{4\sin k_1d_1 \sin k_2d_2} \end{aligned} \tag{6b}$$

The last expression can be easily generalized to relate the atomic displacements in successive periods in superlattices with arbitrary number of layers per period. For an infinite superlattice, Bloch's theorem requires that the displacement vectors $\mathbf{A}_1^{(l)}$ and $\mathbf{A}_1^{(l+1)}$ differ only by a phase factor e^{iqd}, i.e.,

$$\mathbf{A}_1^{(l+1)} = e^{iqd}\mathbf{A}_1^{(l)} \tag{7}$$

where the wave vector q lies in the interval $-\frac{\pi}{d} < q \leq \frac{\pi}{d}$. This interval defines the first Brillouin zone of the superlattice. Since this zone is normally much smaller than the Brillouin zone of a normal crystal and is normally denoted *mini-Brillouin zone*. Comparing equations 6a and 7 we arrive at the following eigenvalue equation:

$$\mathbf{M}\mathbf{A}_1^{(l)} = e^{iqd}\mathbf{A}_1 I^{(l)} \tag{8}$$

which has non-trivial solution only if $det\left[\mathbf{M} - ||e^{iqd}\right] = 0$. The solution of this determinantal equation yields the *dispersion relation* connecting the frequency ω and wave vector q of the vibrational eigenmodes of an infinite superlattice:[7,8]

$$\cos qd = \cos k_1 d_1 \cos k_2 d_2 - \frac{1 - \cos k_1 a \cos k_2 a}{\sin k_1 a \sin k_2 a} \sin k_1 d_1 \sin k_2 d_2 \tag{9}$$

Figure 3 shows dispersion relations calculated using Eq. 9 for linear chains with different number of atoms in the base. The real (q_R) and imaginary (q_I) parts of the wave vector q are displayed by solid and dashed lines, respectively. We recall here that only mode with $q_I = 0$ can exist in an infinite superlattice. In the calculation we assumed a mass ratio $r_m = 1.5$ between the masses of atoms type 1 and 2. This correspond to the mass ratio between silicon and germanium atoms. In each one of the plots the dot-dashed lines are the dispersion relation of linear chains composed only by atoms type 1 or 2 (see Eq. 3). In Fig. 3a the base contain only one atom of type 1 and another of type 2. This corresponds to the dispersion along the [100]-direction of zinc-blend semiconductors such as GaAs and AlAs. The Brillouin zone of the chain is reduced to a half of that of the unstructured materials. The dispersion contains an *acoustic branch*, where the atoms in the base move in the same direction, and a higher frequency *optical branch* , where the atoms in the base oscillate against each other. The two branches are separated by a *frequency gap*, within which there is no vibrational mode. The gap is associated with frequency ranges where the expression at the left hand side of Eq. 9 assumes values greater than one, leading to unacceptable complex values for the wave vector q ($q_I \neq 0$). Physically, vibration modes with frequencies in the gap range are constructively Bragg-reflected at the superlattice interfaces and can therefore not propagate through the structure.

Figure 3b shows the dispersion for a linear chain with a base composed of four atoms, two in each layer. The dispersion is defined within a mini-Brillouin zone with a dimension four times smaller than that of the unstructured materials, but containing four branches. Only the the first of these branches is a true acoustic branch: in the others the atomic movement of the atoms is generally not in the same direction. Note that frequency gaps open now not only at the border of the mini-Brillouin zone, but also in its center.

The dispersion relation of a superlattice containing five atoms of type 1 and five atoms of type 2 is displayed in Fig. 3c. As in the previous case, the Brillouin zone has an extension equal to $2\pi/10(n_1 + n_2)a$, where $n_1 + n_2 = 10$ is the number of atoms in the base. The modes are distributed in dispersion branches separated by small frequency gaps. The number of dispersion branches is equal to the number of atoms in the base.

A deeper insight into the properties of the vibrational modes can be gained by analyzing the atomic displacements. In Fig. 3d. we show the atomic displacement within a period for the modes indicated in the dispersion of Fig. 3c by $\underline{1}$, $\underline{2}$, and $\underline{3}$. All the modes are located in the center of the mini-Brillouin zone ($q_R = 0$) so that the atomic displacement pattern is the same in all periods of the superlattice. The vibration modes (or phonons) can be classified into two categories:

- *Confined phonons*: Let us consider modes 1 and 2 in Fig. 3d. The atomic displacement assumes small values in layer (2) and the atomic motion is practically *confined* to the atoms in layer (1). These modes are located in a frequency range where the dispersion relation of the bulk materials do not overlap in frequency and only material (1) has propagating modes. According to Eq. 3 the bulk wave vector k_2 in layer (2) has an imaginary component. The vibrations cannot propagate in this layer and the atomic displacements decays exponentially. The situation is analogous to that of electrons in a potential well: the vibrational wave function is confined to the phonon well (layer (1)) and decays exponentially in the phonon barrier (layer (2)).

Due to the existence of a phonon well the interaction between vibrations in successive layers of type (1) is very small, and, as a consequence, all modes in a dispersion branch have practically the same branch frequency ω_i^B (the branch index $i = 1, 2, 3...$ is counted from the lower frequency branch). In fact, ω_i^B can be calculated to a very good precision by assuming the atoms in layer (2) *do not* move at all. The situation is similar to that of a rope vibrating with fixed ends. The allowed wave vectors are given by:

$$k_1 a = i\frac{\pi}{n_1+1} \qquad \text{so that} \qquad \omega_i^B = \omega_{max}^{(1)} \sin\left(i\frac{\pi}{(n_1+1)}\right) \tag{10}$$

Note that the first and second dashed curves in Fig. 3d have wavelengths $2\pi/k_1$ obeying the following equation.

- *Folded phonons*: in the frequency range where the dispersion curves of the bulk layer materials overlap the vibration modes can propagate in both materials, such as in case (3) of Fig. 3d. This is the case of acoustic phonons in superlattices. In a first approximation, the dispersion relation in this case is obtained by *folding* into the mini-Brillouin zone of the superlattice the dispersion relation of a fictitious material having vibrational properties lying between those of the layer materials (the dispersion relation of this fictitious material lies between those of the bulk layer materials). In addition, the superlattice dispersion presents frequency gaps in the center and at the boarder of the mini-Brillouin zone. The situation is similar to that of conduction band electrons in a superlattice with energy superior to the barrier: there is no localization and the strong interaction between successive layers leads to the formation of energy bands.

The above remarks remain valid if the layer materials exhibit dispersion relations more complicated than the simple case considered here where the layer materials have just one atom in the unit cell. The classification of the phonons in confined and folded depend in all cases on whether the bulk dispersions of the constituents overlap in frequency or not. Experimental evidence for the existence of both kinds of modes have been extensively reported in the literature.[1,9,10]

3. CONTINUOUS MODELS FOR ACOUSTIC PHONONS

The dispersion relation of low frequency acoustic phonons in a superlattice can be calculated in the elastic continuum approximation where the phonons are well described by elastic sound waves propagating in a periodically structured medium.[11] The dispersion relation for these phonons can therefore be calculated by means of the elasticity theory. This leads to some simplification of the calculation procedure relative to discrete models. In this section we shall calculate the dispersion relation for these phonons, for arbitrary propagation directions in the superlattice and for arbitrary number of layers per period. Throughout the section we will assume that the layer materials are isotropic. The model presented can also be sophisticated to take into account the anisotropy of crystalline materials.

In the continuum limit, each superlattice period of length d (see Fig. 4) can be considered to consist of N layers, each with a thickness d_i, a density ρ_i, and sound velocities v_i^{LA} and v_i^{TA} for longitudinal acoustic (LA) and transverse acoustic (TA) waves, respectively ($i = 1, 2, ...N$). Due to the invariance of the superlattice under rotation around the growth axis (z-direction) the wave vector **q** of the elastic waves can always be assumed to lie in the xz-plane. The eigenmodes for an arbitrary frequency ω and propagation direction φ are classified into two categories:[12]

- A pure transverse acoustic (TA) mode polarized in the y-direction (mode T in Fig. 4);[13]
- Two modes with coupled LA and TA character polarized in the xz-plane (modes K_1 and K_2 in Fig. 4).[14,15] The coupling of these modes takes place because a purely longitudinal (or purely transverse) wave impinging obliquely on an interface generates both longitudinal and transverse transmitted and reflected waves.

Because these two categories of modes have orthogonal polarization directions they do not couple and their dispersion relation can be calculated independently, as we shall see in the following. Before proceeding, we shall first examine the simpler case where the phonons propagate perpendicular to the superlattice layers ($q_x = 0$) in a structure with two layers per period. This corresponds to the situation examined in the previous section. An examination of Fig. 4 shows that in this case the modes K_1

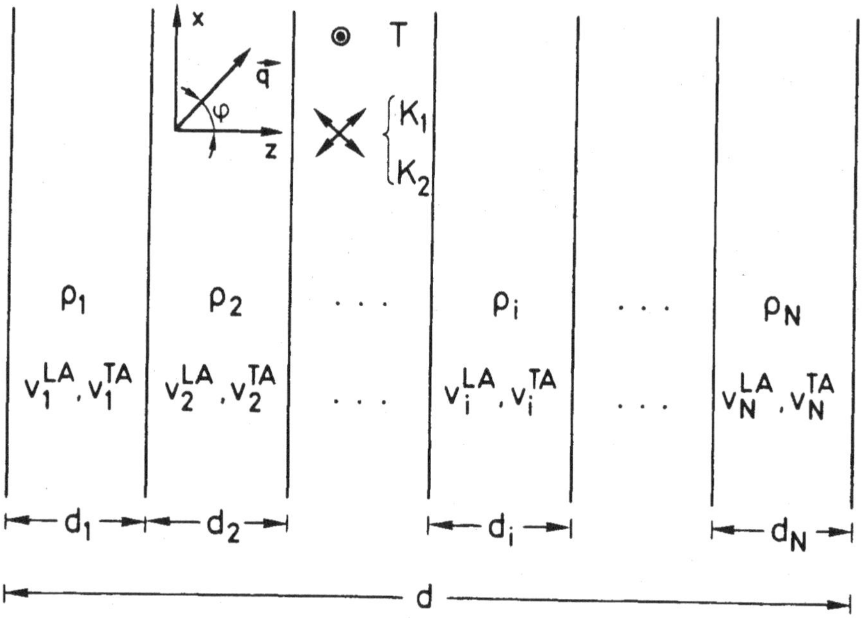

Figure 4: Schematic representation of a period of a superlattice and possible acoustic modes (for details see text).

and K_2 decouple and one of them (K_1) becomes a pure LA wave and the other (K_2) a pure TA wave. Mode K_2 is degenerate with the transverse mode T. The dispersion relations for the TA and LA waves have the same algebraic form and differ only through the different sound velocities for the two modes. It can be obtained directly from Eqs. 3 and 9 by taking the limit $k_1a, k_2a \to 0$. The resulting expression is the well-known Rytov's dispersion relation for acoustic waves in superlattices:[1,16]

$$\cos qd = \cos k_1d_1 \cos k_2d_2 - \frac{1+F^2}{2F} \sin k_1d_1 \sin k_2d_2 \tag{11a}$$

where $F = \rho_1v_1/\rho_2v_2$ is the ratio of the acoustic impedances of the layers and the

k_i's are now given by:

$$k_i = \omega / v_i \tag{11b}$$

This dispersion relation is composed by a series of folded branches separated by frequency gaps. It is completely equivalent to Eq. 9 in the low-frequency region where the dispersion relation of the bulk materials are linear (constant sound velocity, as is shown in Fig. 3c).

3.1 Transverse modes

These vibrations propagate in all superlattice layers with the same wave vector component q_x in the x-direction (see Fig. 4). This is due to the conservation of the parallel component of the wave vector of the acoustic wave upon reflection at or transmission through a plane interface. The displacement field of the acoustic wave $u_j^{(l)}(x, \xi, t)$ in the jth-layer of the lth period is the superposition of two transverse waves with the same wave vector $q_x \hat{\mathbf{e}}_\mathbf{x}$ in the x-direction and opposite wave vectors $\kappa_j^t \hat{\mathbf{e}}_\mathbf{z}$ and $-\kappa_j^t \hat{\mathbf{e}}_\mathbf{z}$ in the z-direction:[13]

$$u_j^{(l)}(x, \xi, t) = e^{ilq_z d} e^{i(q_x x - \omega t)} [A_j^+ e^{i\kappa_j^t \xi} + A_j^- e^{-i\kappa_j^t \xi}] \tag{12a}$$

where ξ is a local z-coordinate relative to the left side of each layer and:

$$\kappa_j^t = \sqrt{q^2 - q_x^2} = \sqrt{(\frac{\omega}{v_j^{TA}})^2 - q_x^2} \tag{12b}$$

The constants A_j^+ and A_j^- $j = 1, 2, \ldots, N$ in Eq. 12a are determined by the boundary conditions at the interfaces, where both the displacement field $u_j^{(l)}$ and the stress $\sigma_j^{(l)} \hat{\mathbf{e}}_\mathbf{x}$ must be continuous functions. For transverse vibrations polarized in the x-direction the only non-zero stress component is

$$\sigma_j^{(l)} \hat{\mathbf{e}}_\mathbf{x} = \sigma_{zx,j}^{(l)} \hat{\mathbf{e}}_\mathbf{x} = c_{44,j}^{(l)} \frac{\partial u_j^{(l)}}{\partial \xi} \hat{\mathbf{e}}_\mathbf{x} \tag{13}$$

The displacement and stress field in each point of the superlattice can be resumed in a vector $\mathbf{U}_j^{(l)}(\xi,\omega,t)$ given by: [13]

$$\mathbf{U}_j^{(l)}(x,\xi,\omega,t) = \begin{pmatrix} u_j^{(l)} \\ i\sigma_j^{(l)} \end{pmatrix} = e^{iq_x x} e^{i(l\kappa_j^t d - \omega t)} \cdot \underbrace{\begin{pmatrix} e^{i\kappa_j^t \xi} & e^{-i\kappa_j^t \xi} \\ \kappa_j^t c_{44,j} e^{i\kappa_j^t \xi} & -\kappa_j^t c_{44,j} e^{-i\kappa_j^t \xi} \end{pmatrix}}_{\mathbf{M}_j(\xi)} \underbrace{\begin{pmatrix} A_j^+ \\ A_j^- \end{pmatrix}}_{\mathbf{A}_j} \tag{14}$$

The continuity of $\mathbf{U}_j^{(l)}(\xi,\omega,t)$ at each interface leads to the expression:

$$\mathbf{U}_{j+1}^{(l)}(0,\omega,t) = \mathbf{U}_j^{(l)}(d_j,\omega,t) \Longrightarrow \mathbf{A}_{j+1} = \mathbf{M}_{j+1}^{-1}(0)\mathbf{M}_j(d_j)\mathbf{A}_j \tag{15}$$

and to the following eigenvalue problem:

$$\mathbf{M}(\omega)\mathbf{A}_1 = e^{iq_z d}\mathbf{A}_1 \tag{16a}$$

where

$$\mathbf{M}(\omega) = \mathbf{M}_1^{-1}(0)\mathbf{M}_N(d_N)\ldots\mathbf{M}_{j+1}^{-1}(0)\mathbf{M}_j(d_j)\ldots\mathbf{M}_2^{-1}(0)\mathbf{M}_1(d_1) \tag{16b}$$

For each frequency ω the factors $e^{iq_z d}$ and therefore the dispersion relation are given by the eigenvalues of the matrix $\mathbf{M}(\omega)$. In the case of superlattices with two layers per period, Eq. 16a leads to an analytical expression equal to Eq. 11a, provided that the impedance ration F and wave vector k_j are substituted by $\rho_1 v_1^{TA}/\rho_2 v_2^{TA}$ and κ_j^t, respectively.

3.2 Coupled LA-TA Modes

The dispersion relation for coupled LA-TA modes was calculated by Hotz and Siems for superlattice structures with isotropic layers.[14,15] In each layer, the eigenmodes are, in this case, the superposition of two transverse and two longitudinal plane waves with the same wave vector component $q_x\hat{e}_x$ along the layers (x-direction). As in Eq. 12 the z-component of the wave vector of the TA-waves within the jth layer is given by:

$$\kappa_j^t = \pm\sqrt{\left(\frac{\omega}{v_j^{TA}}\right)^2 - q_x^2} = \pm\frac{\omega}{v_j^{TA}}\cos(\varphi_j^t) \qquad (17a)$$

and that of the longitudinal waves by:

$$\kappa_j^l = \pm\sqrt{\left(\frac{\omega}{v_j^{LA}}\right)^2 - q_x^2} = \pm\frac{\omega}{v_j^{LA}}\cos(\varphi_j^l) \qquad (17b)$$

where φ_j^l and φ_j^t are the angles between the z-axis and the propagation direction of the LA- and TA- waves in the jth, respectively. If $A_{l,j}^+$ ($A_{l,j}^-$) denotes the amplitude of the LA plane waves in the jth layer with wave vector component κ_j^l in the positive (negative) z-direction and $A_{t,j}^+$ ($A_{t,j}^-$) the corresponding amplitude of the TA plane wave, then the displacement field in the jth layer of the lth period can be written in the following way:

$$\mathbf{u}_j^{(l)}(\xi, x, \omega, t) = \begin{pmatrix} u_{z,j}^{(l)}(x,\xi,t) \\ u_{x,j}^{(l)}(x,\xi,t) \end{pmatrix} = e^{ilq_zd}e^{i(q_xx-\omega t)} \cdot$$

$$\begin{pmatrix} -iC_j^l(\xi)C_{\varphi,j}^l & iS_j^l(\xi)C_{\varphi,j}^l & -iS_j^t(\xi)S_{\varphi,j}^t & iC_j^t(\xi)S_{\varphi,j}^t \\ S_j^l(\xi)S_{\varphi,j}^l & C_j^l(\xi)S_{\varphi,j}^l & C_j^t(\xi)C_{\varphi,j}^t & S_j^t(\xi)C_{\varphi,j}^t \end{pmatrix} \underbrace{\begin{pmatrix} i(A_{l,j}^+ - A_{l,j}^-) \\ A_{l,j}^+ + A_{l,j}^- \\ A_{t,j}^+ + A_{t,j}^- \\ i(A_{t,j}^+ - A_{t,j}^-) \end{pmatrix}}_{\mathbf{A}_j} \qquad (18a)$$

where

$$C_j^{l,t} = \cos(\kappa_j^{l,t}\xi), \qquad S_j^{l,t} = \sin(\kappa_j^{l,t}\xi) \qquad (18b)$$

and

$$C_{\varphi,j}^{l,t} = \cos(\varphi_j^{l,t}), \qquad S_{\varphi,j}^{l,t} = \sin(\varphi_j^{l,t}). \qquad (18c)$$

ξ is a local z-coordinate with respect to the left interface of each layer and $\mathbf{A}_j$ is a four component vector containing the amplitude information of the wave.

At the interface of two adjacent layers both $u_{x,j}^{(l)}$, $u_{z,j}^{(l)}$, and the stress components $\sigma_{zz,j}^{(l)}\hat{\mathbf{e}}_z = c_{11,j}^{(l)}\frac{\partial u_z^{(l)}}{\partial z} + c_{12,j}\frac{\partial u_x^{(l)}}{\partial x}\hat{\mathbf{e}}_z$ and $\sigma_{xz,j}\hat{\mathbf{e}}_x = \frac{c_{44,j}}{2}[\frac{\partial u_z}{\partial x} + \frac{\partial u_x}{\partial z}]\hat{\mathbf{e}}_x$ are continuous. As in Eq. 14, these conditions can be condensed in a vector $\mathbf{U}_j^{(l)}(\xi, x, \omega, t)$ given by:

$$\mathbf{U}_j^{(l)}(\xi, x, \omega, t) = \begin{pmatrix} u_{z,j}^{(l)} \\ u_{x,j}^{(l)} \\ \sigma_{zz,j}^{(l)} \\ 2\sigma_{xz,j}^{(l)} \end{pmatrix} = e^{ilq_z d} e^{i(q_x x - \omega t)} \mathbf{H}_j \mathbf{L}_j(\xi) \mathbf{A}_j \tag{19a}$$

where

$$\mathbf{H}_j = \begin{pmatrix} -iC_{\varphi,j}^l & 0 & & iS_{\varphi,j}^t \\ & S_{\varphi,j}^l & C_{\varphi,j}^t & 0 \\ 0 & iBS_{\varphi,j}^l & -iCC_{\varphi,j}^t & 0 \\ CC_{\varphi,j}^l & 0 & 0 & BS_{\varphi,j}^t \end{pmatrix} \tag{19b}$$

with $B = q_x(\rho_j \omega^2/q_x^2 - 2c_{44,j})$, $C = 2q_x c_{44,j}$, and

$$\mathbf{L}_j(\xi) = \begin{pmatrix} C_j^l(\xi) & -S_j^l(\xi) & 0 & 0 \\ S_j^l(\xi) & C_j^l(\xi) & 0 & 0 \\ 0 & 0 & C_j^t(\xi) & S_j^t(\xi) \\ 0 & 0 & -S_j^t(\xi) & C_j^t(\xi) \end{pmatrix} \tag{19c}$$

The physical meaning of the of the matrices $\mathbf{H}_j$ and $\mathbf{L}_j(\xi)$ is as follows: the product $\mathbf{H}_j\mathbf{A}_j$ establishes the boundary condition vector $\mathbf{U}_j^{(l)}$ at the left hand side of each layer ($\xi = 0$). The matrix $\mathbf{L}_j(\xi)$ translate the z-coordinate to the point ξ. The product $\mathbf{H}_j\mathbf{L}_j(\xi)\mathbf{A}_j$ gives therefore the boundary condition vector at the point ξ.

From the continuity of the function $\mathbf{U}_j^{(l)}$ at the interfaces follows the following relationship:

$$\mathbf{U}_j^{(l)}(d_j, x, \omega, t) = \mathbf{U}_j^{(l)}(0, x, \omega, t) \Longrightarrow \mathbf{A}_{j+1} = \mathbf{H}_{j+1}^{-1}\mathbf{H}_j\mathbf{L}_j(d_j)\mathbf{A}_j \tag{20a}$$

where

$$\mathbf{A}_{N+1} = \mathbf{M}(\omega)\mathbf{A}_1 \tag{20b}$$

with

$$\mathbf{M}(\omega) = \mathbf{H}_{N+1}^{-1}\mathbf{H}_N\mathbf{L}_N(d_N)\mathbf{H}_N^{-1}\ldots\mathbf{H}_j\mathbf{L}_j(d_j)\mathbf{H}_j^{-1}\ldots\mathbf{H}_2^{-1}\mathbf{H}_1\mathbf{L}_1(d_1) \tag{20c}$$

where $\mathbf{H}_{N+1} = \mathbf{H}_1$. We obtain therefore the following eigenvalue equation for the displacement vector $\mathbf{A}_1$:

$$\mathbf{M}(\omega)\mathbf{A}_1 = e^{iq_z d}\mathbf{A}_1 \tag{21}$$

3.3 Internal gaps

Figure 5 shows a numerical solution of Eq. 21 for a structure with two layers per period with $\rho_1 = \rho_2 = 2.32\frac{g}{cm^3}$, $d_1 = 36\mathring{A}$, $d_2 = 51\mathring{A}$, $v_1^{LA} = 9.6 \times 10^5\frac{cm}{s}$, $v_2^{LA} = 7.1 \times 10^5\frac{cm}{s}$ and $v_j^{TA} = 0.65v_j^{LA}$, $j = 1, 2$. These are typical values for amorphous glow discharge a-Si:H/a-SiN$_x$:H superlattices.[17] The calculations were performed for different values of q_x. For $q_x = 0$ (solid line), corresponding to propagation perpendicular to the layers, the modes with LA and TA character decouple and the dispersion is the superposition of the pure LA and TA dispersions. They have frequency gaps in the center ($q_z = 0$) and at the boundary ($q_z = \frac{\pi}{d}$) of the mini-Brillouin-Zone. For $q_x \neq 0$ the modes have mixed LA and TA characters. The dispersion shows additional gaps *within* the mini-Brillouin Zone, at the points where the pure LA and the pure TA dispersions cross. These additional gaps are called *internal* gaps. As for the normal gaps, they are frequency ranges where the z-component of the wave vector has an imaginary part, so that the elastic waves cannot propagate in the z-direction.

From the view point of the perturbation theory the frequency gaps for elastic waves in superlattices are due to the lift of degeneracy of modes at the points where two (unperturbed) dispersion branches cross. For propagation in the z-direction the layer have in the unperturbed case the same acoustic impedance. In this case is valid the picture of folded phonons without gaps in the center and at the boundary of the mini-Brillouin Zone. The perturbation is introduced by the periodic modulation of the acoustic properties. Gaps open in the center and at the boundary of the mini-brillouin Zone where the folded LA or folded TA branches cross. At these points the interbranch interaction is strongest (vanishing energy denominators) The degeneracy of modes at the cross points of the LA and TA branches is not lifted because the modes have different polarization directions and can therefore not interact with each other. For non-vanishing wave vectors $q_x \neq 0$, however, the modes have mixed LA and TA character and can interact, leading to the opening of the internal gaps.

Away from the internal gaps the mixing of the LA and TA character of the modes is relatively small. For each frequency, one mode has a predominant LA character (LA-like mode) and the other a predominant TA-character(TA-like mode). Near the internal gaps, on the other hand, a impinging LA-like wave (TA-like wave) on the oblique superlattice interfaces generates mainly TA-like (LA-like) transmitted and reflected waves.[21] The mode mixing increases strongly near the internal gaps and the modes can no longer be classified as LA-like or TA-like.

The central frequency of the ith internal gap is denoted by $\bar{\omega}_i^{LA-TA}$. Both the central frequency and the width of the internal gap depend on the magnitude of q_x (see Fig. 5). For small q_x ($q_x << \frac{\pi}{d}$) the central frequency $\bar{\omega}_i^{LA-TA}$ corresponds to the crossing frequency $\omega_{\pm i}^x$ of the pure LA and Ta dispersions. In the linear range of these dispersions, i.e., away from the center and from the boundary of the mini-Brillouin Zone, the crossing frequency ω_i^x is given the following expression:

$$\omega_i^x = \frac{2}{1+r_v} i\bar{\omega}_1^{b,TA}, \qquad l = 1, 2, 3, \ldots \tag{22a}$$

where

$$r_v = \frac{v_{eff}^{TA}}{v_{eff}^{LA}} = \frac{\bar{\omega}_1^{b,TA}}{\bar{\omega}_1^{b,LA}} \tag{22b}$$

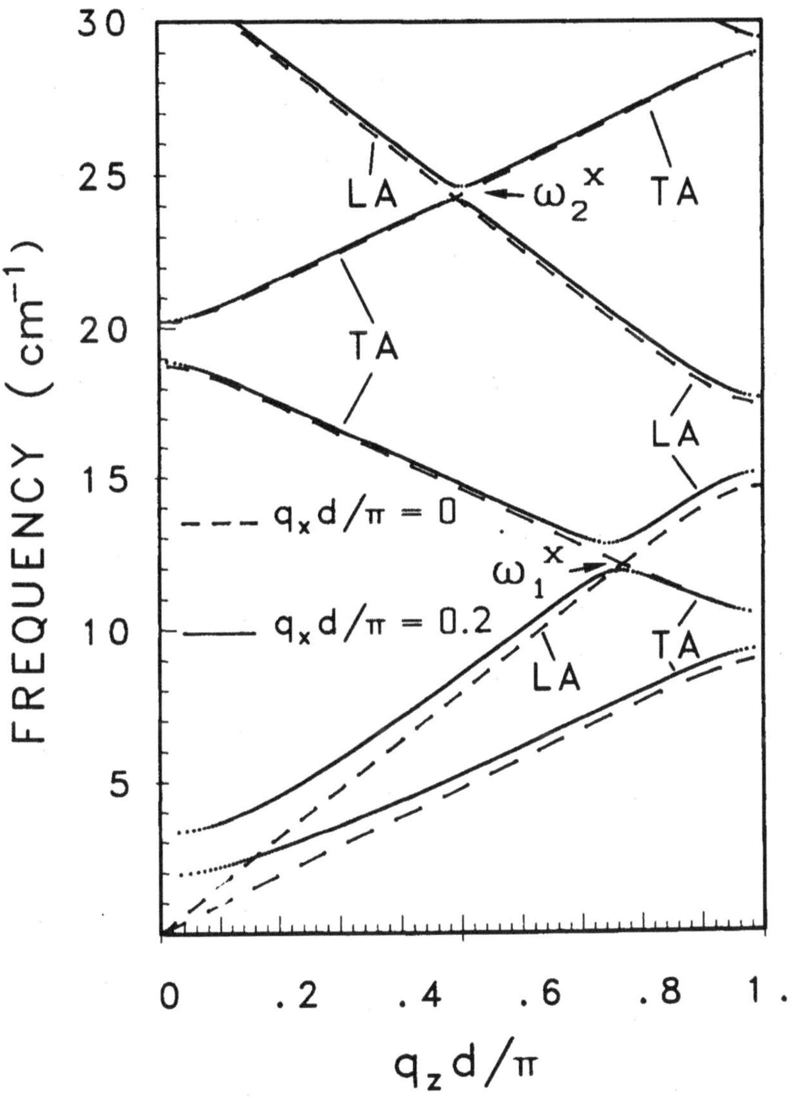

Figure 5: Projection of the dispersions relation in the (ω, q_z)-plane for mixed LA-TA modes with wave vector $(q_x, 0, q_z)$. The parameter used in the calculations were:: $\rho_1 = \rho_2 = 2.32\frac{g}{cm^3}$, $d_1 = 36\AA$, $d_2 = 51\AA$, $v_1^{LA} = 9.6 \times 10^5\frac{cm}{s}$, $v_2^{LA} = 7.1 \times 10^5\frac{cm}{s}$ and $v_j^{TA} = 0.65 v_j^{LA}$, $j = 1, 2$.

is the ratio of the effective sound velocity for TA and LA waves in the z-direction. The frequencies $i\bar{\omega}_1^{b,LA}$ and $i\bar{\omega}_1^{b,TA}$ are, respectively, the central frequency of the ith LA and TA gaps for propagation perpendicular to the layers.

4. PHONON TRANSMISSION THROUGH SUPERLATTICES

Recently, a number of experiments have been reported where the transmission of high-frequency acoustic phonons through multilayer structures is measured experimentally.[17-22] In these experiments, a phonon beam impinges on the surface of the multilayer and the transmitted phonon intensity on the other side of the structure is measured as a function of the frequency.[23] The phonon transmission can be calculated for structures with arbitrary number of layers using the results of the last section (see Eq. 20b). As an example, we shall consider the case where the incident wave in the first layer (layer 1) is a pure transverse wave polarized in the xz-plane (see Fig. 4), so that only mixed LA-TA modes will be excited. This wave has an amplitude T_1^+ and has a positive z-wave vector component. We denote by T_1^- and by L_1^- the amplitude of the reflected TA and LA waves in the first layer, respectively (The superscripts + and − denote propagation in the positive and in the negative z-direction, respectively). According to Eq. 16, the amplitude vector $\mathbf{A}_1$ in this layer is given by:

$$\mathbf{A}_1 = \begin{pmatrix} -iL_1^- \\ L_1^- \\ T_1^+ + T_1^- \\ i(T_1^+ - T_1^-) \end{pmatrix} \tag{23a}$$

In general, the transmitted wave in the last layer (layer $N+1$) layer is the superposition of a TA wave with amplitude T_{N+1}^+ and a LA wave of amplitude L_{N+1}^+. These two waves have wave vector components in the positive z-direction. The amplitude vector $\mathbf{A}_{N+1}$ in the last layer is then given by the following expression:

$$\mathbf{A}_{N+1} = \begin{pmatrix} iL^{+}_{N+1} \\ L^{+}_{N+1} \\ T^{+}_{N1+1} \\ iT^{+}_{N+1} \end{pmatrix} \tag{23b}$$

The unknown amplitudes L_1^-, T_1^-, L^+_{N+1} and T^+_{N+1} follow from the solution of the linear system of equations expressed by Eq. 20b. The transmission through the structure, defined as the ration between the transmitted and incident acoustic power, is then given by:

$$T = \frac{\rho_{N+1} v^{TA}_{N+1} \left| T^{+}_{N+1} \right|^2 + \rho_{N+1} v^{LA}_{N+1} \left| L^{+}_{N+1} \right|^2}{\rho_1 v_1^{TA} \left| T_1^{+} \right|^2} \tag{24}$$

As an example, we show in Fig. 6 the calculated transmission spectrum for a glow discharge a-Si:H/a-SiN$_x$:H superlattice with 20 periods, each containing a silicon layer 47 Å thick and a silicon nitrite layer 36 Å thick.[12] The structure is grown on a crystalline silicon substrate. The incident TA phonons are polarized in the plane perpendicular to the superlattice layers and travel towards the crystalline silicon-superlattice interface with an angle of incidence ϕ. The transmitted acoustic power was calculated at the opposite surface of the superlattice. Curve (a) is for normal phonon incidence ($\phi = 0$): it shows only the TA frequency gaps at the boundary and in the center of the mini-Brillouin-zone. When ϕ deviates slightly from zero ($\phi = 0.1$ rad $\sim 6°$ in Fig. 6) the internal gaps appear in the spectrum.

5. CONCLUSIONS

In this paper we reviewed the basics concepts related to phonons in superlattice structures. We introduced the concepts of phonon folding and phonon confinement using a simple linear chain model for the superlattices. In the second part of the paper we discussed continuous model for acoustic phonon propagation in super-

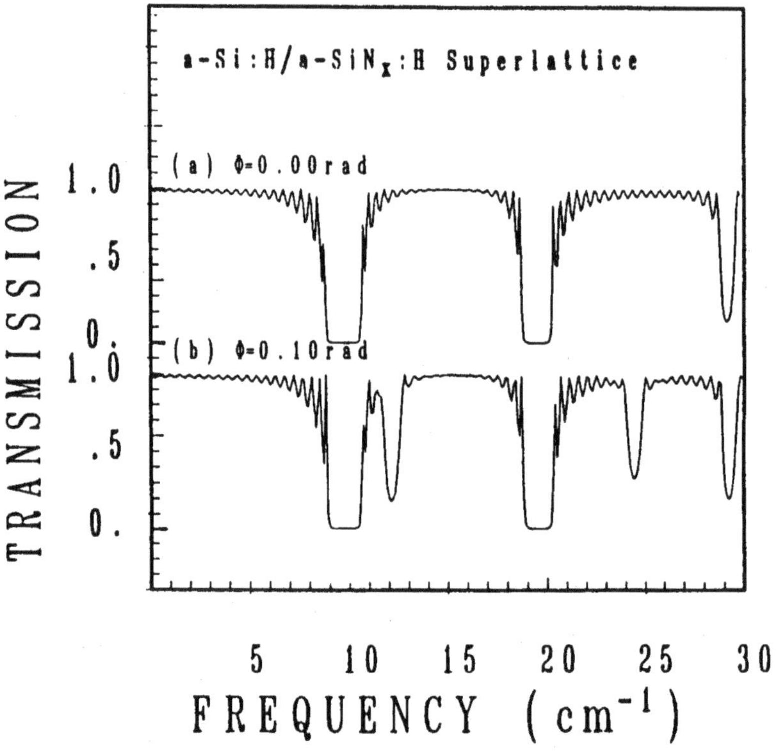

Figure 6: Calculated phonon transmission spectra for the superlattice structure of Fig. 5. The curves are for (a) perpendicular phonon incidence ($\phi = 0$) (b) $\phi = 0.1$.

lattices. Emphasis was given to calculations of the dispersion relation for arbitrary polarization and propagation directions of long wavelength acoustic phonons.

ACKNOWLEDGEMENTS: This work was partially supported by CNPq, Brazil and by FAPESP, São Paulo, Brazil.

References

[1] Colvard, C., Grant, T.A., Klein, M. V., Merlin, R., Fischer, R., Morkoc, H., and Gossard, A.C, *Phys. Rev. B* **31**, 2080 (1985).

[2] Klein, M.V., *IEEE J. Quantum Electron* **QE-22**, 1760 (1986).

[3] M. Cardona, in *Lectures on Surface Science*, Eds.: Castro, G.R., and Cardona, M. (Springer-Verlag, Heidelberg, 1987), p. 2.

[4] Molinari, E. and Fasolino, A., *Superlattice and Microstructures* **4**, 449 (1988).

[5] Santos, P.V. and Ley, L., *Superlattice and Microstructures* **5**, 43 (1988).

[6] Cardona, M., *Superlattices and Microstructures* **5**, 27 (1988).

[7] Jusserand, B. and Cardona, M., in *Light Scattering in Solids V*, Eds.: Cardona, M. and Güntherrodt, G. (Spriger-Verlag, Berlin, 1988), p. 49.

[8] Albuquerque, E.L., Fulco, P., and Tilley, D.R., *Revista Brasileira de Física* **16**, 315 (1986).

[9] Colvard, C., Merlin, R., Klein, M.V. and Gossard, A.C. , *Phys. Rev. Lett.* **45**, 298 (1980).

[10] Sood, A.J., Menendez, J., and Cardona, M., *Phys. Rev. Lett.* **54**, 2111 (1985).

[11] Brekhovskikh, L.M., *Waves in Layered Media* (Academic Press, New York, 1960).

[12] Santos, P.V., PhD Dissertation, Stuttgart, 1988.

[13] Camley, R.E., Djafari-Rouani, B. , Dobryznski, L. and Maradudin, A.A, *Phys. Rev. B* **27**, 7318 (1983).

[14] Hotz, R. and Siems, R., *Solid State Comm.* **51**, 793 (1984).

[15] Hotz, R. and Siems, R., in *Recent Developments in Condensed Matter Physics vol 4*, Eds.: Devreese, J.T., Lemmens, L.F., van Doren, V.E., and van Royen, J. (Plenum Press, New York, 1981), p. 167.

[16] Rytov, S.M., *Akust.. Zh.* **2**, 71 (1956) [*Sov. Phys. Acoust.* **2**, 68 (1956)].

[17] Santos, P.V., Ley, L., Mebert, J., and Koblinger, O., *Phys. Rev. B* **36**, 4858 (1987).

[18] Narayanamurti, V., Stormer, H.L., Chen, M.A., Gossard, A.C. and Wiegmann, W., *Phys. Rev. Lett.* **43**, 2012 (1979).

[19] Koblinger, O., Mebert, J., Dittrich, E., Döttinger, S., Eisenmenger, W., Santos. P.V. and Ley, L., *Phys. Rev. B* **35**, 9372 (1987).

[20] Santos, P.V., Ley, L., Mebert, J. and Koblinger, O., *Phys. Rev. B* **36**, 1306 (1987).

[21] Tamura, S. and Wolfe, J.P., *Phys. Rev. B* **35**, 2528 (1987).

[22] Hurley, D.C, Tamura, S., Wolfe, J.P. und Morkoc, H., *Phys. Rev. Lett.* **23**, 2446 (1987).

[23] Eisenmenger, W., in *Physical Acoustics, vol XII*, Eds.: Mason, W.A. and Thurson, R.N. (Academic Press, New York, 1979), p. 79.

TRANSITION METAL IMPURITIES IN SEMICONDUCTORS

A. Fazzio
Instituto de Física, Universidade de São Paulo,
Caixa Postal 20516, 01498 São Paulo, SP, Brazil

ABSTRACT

A brief review of the theoretical study of transition metal impurities in semiconductors is presented. Using a Multiplet method we discussed three problems: (i) $GaAs:Cr^{2+}$ under pressure, (ii) 4d and 5d series transition metal impurities and (iii) band-edge discontinuities relative to 3d impurities-levels.

1. INTRODUCTION

The transition metal 3d (TM) impurities in semiconductors constitute a distinct class of defects that induce a strong disturbance in the covalent bonding of the host lattice, introducing deep levels in the gap region. The microscopic structure of TM in semiconductors has been clarified through several experimental works[1] and theoretical investigations including multiplet corrections[2,3,4]. In this work I discuss some particular aspects of TM in III-V and II-VI semiconductors: (i) the system $GaAs:Cr^{2+}$ under hydrostatic pressure; (ii) the 4d and 5d impurity series; and (iii) possible guide to prediction of band-edge discontinuites using acceptor and donor states of TM impurities.

2. SOME IMPORTANT ASPECTS OF THE TM IMPURITIES PROBLEM

The TM impurities in III-V and II-VI semiconductors occupy preferentially the substitutional site, replacing the cation. The d-orbital of the impurities in the tetrahedral crystal field symmetry should split in e and t_2 symmetry orbitals. The acceptor and donor ionization energies can be of the order of 0.1 - 2.5 eV, the Mott-Hubbard energies are typically of the order of 0.2 - 2.0 eV, intra-d excitations are seen to be of the order of 0.3 - 2.5 eV, and crystal field splittings range from 0.3 to 1.5 eV. It is important to note here that all relevant energies can fit into the typical energy range of a semiconductor gap; in other words the TM impurity has a substantial electronic interaction with the host lattice[3]. While this interaction can be seen as resulting from strong hybridization of the impurity d-states, it is also clear that d-related multiplet splittings have to be the consequence of strong localization or correlation effects. In the next section I give a brief presentation of the multiplet model I will use in the applications.

3. MULTIPLET MODEL

We may write the hamiltonian for the N-electron system as a sum $\hat{H} = \hat{H}_0 + \hat{H}_1$, where $\hat{H}_0$ stands for the bare-ion (one-electron) term, and $\hat{H}_1$ the electron-electron term. If the N electrons are to be distributed among e and t_2 symmetry orbitals we may characterize a mean-field configuration by the occupancy numbers (m,n) of the states.

The interelectronic interaction $\hat{H}_1$ splits the total energy of each single configuration (SC), say, $e^m t^n$, by an amount $E_{SC}(m,n)$. It also introduces a configuration mixing (CM) interaction energy $E_{CM}(m,n;m',n')$ between terms of the same space and spin symmetry. In mean-field electronic structure calculations (which are done for one fixed configuration at a time), there is no place for configuration mixing. However, since the totally-symmetric (a_1) component of the charge density with its attendant interelectronic interaction $\langle \hat{H}_1 \rangle$ is retained, such calculations incorporate the corresponding <u>average</u> of all single configuration energies for the particular configuration for which the calculation is done. This average SC energy $\hat{E}(m,n)$ can be written as a simple weighted sum

$$\hat{E}(m,n) = \sum_i \omega_i \, E^i_{SC}(m,n) \, , \tag{1}$$

where the weights

$$\omega_i = (2S+1)\, g_\Gamma \Big/ \sum_{S,\Gamma} (2S+1)\, g_\Gamma$$

include both spin (S) and space (Γ) degeneracies. We can now express the SC energy $E^i_{SC}(m,n)$ relative to the mean-field average $\hat{E}(m,n)$, defining thereby the single-configuration energy shift $\Delta E^{i}_{SC}(m,n)$ which constitutes a correction to MF theory. The diagonal element of the $\hat{H}_0 + \hat{H}_1$ matrix is then

$$\Gamma_{D_{\alpha\alpha}}(m,n) = \hat{E}(m,n) + \Delta E^{\alpha}_{SC}(m,n) + k_{\alpha\alpha}\Delta_{CF} \,, \tag{2}$$

where Δ_{CF} is the bare-ion crystal-field splitting, and $k_{\alpha\alpha}$ is an integer, as in the Tanabe-Sugano model. If we measure the average energy of the configurations (m,n) relative to a reference configuration $(m^o n^o)$, defining thereby Δ as

$$\hat{E}(m,n) \equiv \hat{E}(m^o,n^o) + \Delta(m,n;m^o n^o) \,, \tag{3}$$

Eq.(2) can be rewritten as

$$\Gamma_{D_{\alpha\alpha}}(m,n) = \hat{E}(m^o,n^o) + \Delta E^{\alpha}_{SC}(m,n) + \left[\Delta(m,n;m^o n^o) + k_{\alpha\alpha}\Delta_{CF}\right] . \tag{4}$$

The term in brackets in Eq.(4) is the effective crystal-field splitting $\Delta_{eff}(m,n)$ and represents the separation between the total energies of the configuration (m,n) and $(m^o n^o)$, including both bare-ion (Δ_{CF}) and average multiplet contributions (Δ). It equals thus the difference in total energies $E_T(m,n) - E_T(m^o,n^o)$ of MF calculations for the two configurations. The diagonal and nondiagonal elements of the interaction matrices are hence given, respectively, as

$$\Gamma_{D_{\alpha\alpha}}(m,n) = \Delta E^{\alpha}_{SC}(m,n) + \Delta_{eff}(m,n;m^o n^o)$$

and (5)

$$\Gamma_{D_{\alpha\beta}}(m,n;m'n') = E^{\alpha\beta}_{CM}(m,n;m'n') \,.$$

These elements depend on the 10 independent Coulomb integrals between the 3-fold degenerate t_2 orbitals, (ξ,η,ζ) and the 2-fold degenerate e orbitals (ε,θ). Consider the (unknown) transition-atom-centered impurity Wannier orbitals $|e\rangle$ and $|t_2\rangle$ that transform in the limit of a separated crystal like the e and t_2 representations, respectively. We can define the orbital deformation parameters

$$\lambda_e^4 = \langle ee\,|\,ee\rangle_S / \langle dd\,|\,dd\rangle_I \,; \quad \lambda_t^4 = \langle tt\,|\,tt\rangle_S / \langle dd\,|\,dd\rangle_I \,, \tag{6}$$

that measure the ratio between the interelectronic interactions in the solid (S) to those in the free ion (I). For simplicity, assume that $\lambda_{et} = (\lambda_t\lambda_e)^{1/2}$. Here $\langle ee\,|\,ee\rangle$ and $\langle tt\,|\,tt\rangle$ are interelectronic integrals involving the ϵ, Θ and ξ,η,ζ, partners orbitals of the e and t_2 Wannier local orbitals, respectively, and $\langle dd \mid dd \rangle$ are atomic interelectronic integrals involving the pure-d states of the free ion. ΔE^{α}_{SC} and $E^{\alpha\beta}_{CM}$ can be expressed in a standard form in terms of the Wannier function deformation parameters λ_e and λ_t and the Racah parameters B_o and C_o of the free ions. Note that the Racah parameter A, which is configuration dependent, does not appear explicitly in Eq. (5), since the dependence of the multiplet splitting on A(m,n) is transformed into $\Delta_{eff}(m,n;m^o n^o)$. There is no need, therefore, to neglect A, or to approximate it by a constant. Given the free-ion values B_o and C_o, the multiplet corrections to the mean-field energies can be expressed therefore solely in terms of λ_e, λ_t, and Δ_{eff}. Notice that λ_e and λ_t reflect coulomb distortions in both bound and resonant states relative to a free-ion reference system.

4. APPLICATIONS

4.1. Cr^{2+} Luminescence under Hydrostatic Pressure

The internal luminescence of $Cr^{2+}(3d^4)$ in GaAs ($^5E \rightarrow {}^5T_2$) can be observed when the ^5E level is pushed below the conduction band minimum, which is possible through alloying GaAs with AℓAs, or by applying hydrostatic pressure to GaAs. The behavior of this internal transitions under high pressure (up to 4 GPa) and low temperatures has show a puzzling strong quenching of the luminescence. This phenomenon has been tentatively interpreted by Zigone et al[5] as a crossing of the ^5E state with the $^3T_1$ and/or $^1A_1$ states, so that the first excited state would show low spin. We have used the Multiplet model (presented in section 3) for interpretations of those experiments[3]. In our calculation[6] we first determine the parameters λ_e, λ_t and Δ_{eff} to results in accordance with experimental values for observed absorption at zero pressure[7] (λ_e = = 0.9487, λ_t = 0.8165 and Δ_{eff} = 0.401 eV)[8]. Now considering that the t_2 orbitals interact with the first neighbor shell and e orbitals with the second one, we are assuming that t_2 (d_{xy}, d_{xz}, d_{yz}) orbitals are much more affected by compression than e ($d_{x^2-y^2}$, d_{z^2}). Then, we can assume that $d\lambda_e/dP$ is equal to zero, then $d\Delta E/dP$ only depends on $d\lambda_t/dP$ and $d\Delta_{eff}/dP$, resulting

$$\frac{d\Delta E(^5E \rightarrow {}^5T_2)}{dP} = \left[-24\,\lambda_t^3 + 2\,\lambda_e^2\lambda_t \right] B_o \frac{d\lambda_t}{dP} +$$

$$+ \left[-8\,\lambda_t^3 + \lambda_e^2\lambda_t \right] C_o \frac{d\lambda_t}{dP} + \frac{d\Delta_{eff}}{dP} \quad . \tag{7}$$

In our case, where covalence is important, we assume a dependence for Δ_{eff} as $\lambda_t^{3/2}\lambda_e^{5/2}/R^{-5}$ in accordance with the involved transition $e^2t^2 \rightarrow e^1t^2$ in one electron picture. Then

$$\frac{d\Delta_{eff}}{dP} = \left[\frac{5}{2\lambda_t}\frac{d\lambda_t}{dP} + \frac{5}{3} K \right] \Delta^o_{eff} \tag{8}$$

where Δ^o_{eff} is the Δ_{eff} in the absence of hydrostatic pressure and K is the compressibility[7].

Inserting equations (8) in (7) we are able to write $d\Delta E/dP$ depending only on $d\lambda_t/dP$. Now using the experimental value for the dependence of the energy for $^5E \rightarrow {}^5T_2$, that is equal 12 meV/GPa , we obtain $d\lambda_t/dP$, and which can then be used for all multiplet transition.

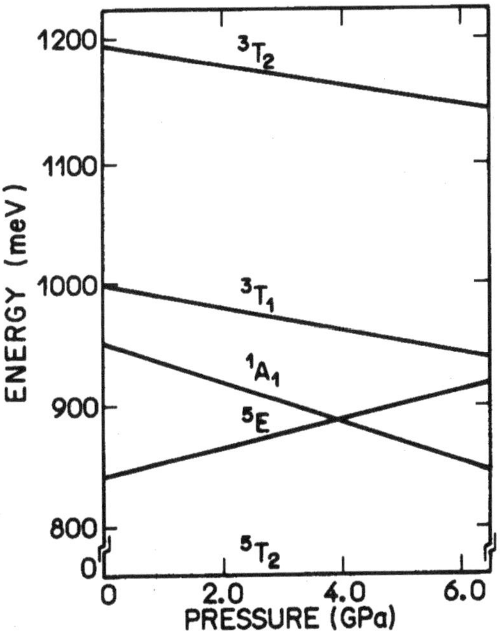

Figure 1. Variation with pressure of the energies of fundamental and excited states associated to the internal transition of Cr^{2+} .

The results are presented in figure 1 for ^5E, $^1A_1$, $^3T_1$ and $^3T_2$ states relative to the ground states $^5T_2$. As we can see, around 3.9 GPa there is a crossing of ^5E and $^1A_1$ states, in very good agreement with experimental results (~ 4.0 GPa), with $d\Delta E(^1A_1 \rightarrow {}^5T_2)/dP \sim -18.0$ meV/GPa. In conclusion, our results agree with the experimental results obtained by Zigone et al[5], and we can note that the $^3T_1$ and $^3T_2$ states is not involved in the crossing.

4.2. 4d and 5d impurities

We investigate the behaviour of the impurities Zr to Rh (4d) and Hf to Ir (5d) in GaAs and GaP using the multiplet method coupled to the Multiple Scattering-Xα one electron method. The experimental data available on these center are still very scarce however these impurities apparently present a much lower solubility limit in III-V compounds than 3d metals; on the other hand they may also show a smaller diffusivity, which can be particularly useful for the growth of substrate materials. In figure 2 I show the mean field (MS-Xα)[9] results for neutral impurities in GaAs for 3d, 4d and 5d series; where only relevant eigen-

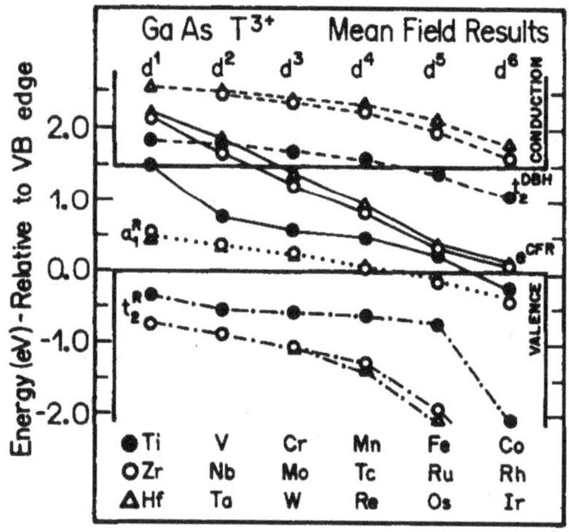

Figure 2. Results obtained through the MS-Xα cluster model for metal impurities in GaAs : solid circles, 3d impurities Ti to Co; open circles, 4d impurities Zr to Rh; and open triangles, 5d impurities Hf to Ir. Only relevant eigenvalues are shown, joined by lines to guide the eye.

values are shown. As I am interested in the determination of ionization level in the gap, acceptors or donors, for 4d and 5d centers, so we must first estimate the multiplet correction affecting mean field results for these transitions. As I mentioned, there are very few experimental data available on intra-d transitions for these impurity centers[10]. We estimate the three mean field parameters needed for multiplet scheme (Δ_{eff}, λ_e and λ_t) from the corresponding 3d impurities. This procedure is justified by the strong similarities in mean-field results for the three series. We may expect slightly larger Δ_{eff} and smaller λ_e and λ_t for the 4d and 5d impurities compared to 3d impurities; as these alterations tend to diminish multiplet effects, we will retain the same parameters for 4d and 5d for 3d centers, which will give us the upper bound to many-electron effects. In figure 3 we have the multiplet corrections for 3d and 5d impurities in the negative charge state TM^{2+} ground state, and corrections for acceptor and donor transitions. The corrections are obtained using the Racah parameters B_o and C_o for the free ion from reference 11. Focusing on acceptor transition, figure 3b, we note that the transition usually retain their characteristics. In the case of W we have an anomaly with the acceptor state W^{2+} switching to low spin, and possibly the neutral state also. In this case the ionization (electron capture) should occur at the e level ${}^4T_1(e^2t^1) \rightarrow {}^1A_1(e^4)$ or ${}^2E(e^3) \rightarrow {}^1A_1(e^4)$ — instead of at the t_2 level as in the case of Cr, where the transition is ${}^4T_1(e^2t^1) \rightarrow {}^5T_2(e^2t^2)$. Then, coupling the information of multiplet correction with mean field results[9] we may give relevant information about ionization levels of 4d and 5d impurities: (i) The impurities Ir and Rh will introduce acceptor levels close to that of Co and no donors, in the gap of GaAs and GaP. (ii) The acceptor levels of Os and Ru will likewise be close to that of Fe; the donor transitions, on the other hand, involve the t_2 state ($\sim$0.5 eV higher for the heavier impurities) and hence these impurities should introduce donor levels easily detectable in both materials. (iii) The large stabilization energy of the ${}^6A_1$ state, coupled to the expected lattice relaxation energy around a symmetric state, should lower the acceptor level of Re into the upper half of the band gap of GaP, and for GaAs the level could appear close to midgap. (iv) In the case of Mo and W the donor level should appear in the upper third of the gap.

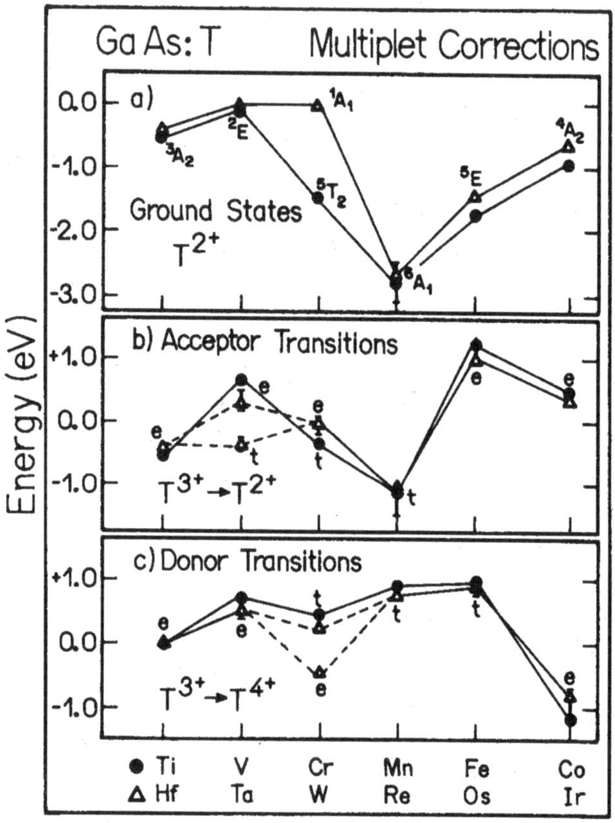

Figure 3. Multiplet corrections for transition metal impurities in GaAs. Circles, 3d impurities Ti to Co; triangles, 5d impurities Hf to Ir. a) corrections to the ground state (indicated) of negatively charged impurities T^{2+}; b) corrections to acceptor transitions, as measured from the valence band; and c) corrections to donor transitions, as measured from the conduction band.

(v) A donor level should also be visible in the upper half of the gap for Nb and Ta in GaAs and GaP, and no acceptors: any detected luminescence[10] should most probably be ascribed to the impurities in the 3+ oxidation state.

4.3. Band-edge Discontinuities

In our systematic analysis of TM impurities in II-VI and III-V compounds, we have found a remarkable similarity, within the same class of compounds, in the impurity interaction with the host lattice. We have, for instance, almost identical hybridization parameters λ_e and λ_t and very similar effective splittings Δ_{eff} for an impurity in GaAs, GaP, or InP or in ZnS, ZnSe, etc. Accordingly, intra-d excitation transitions are also very similar, as are multiplet corrections in general, including those for ionization transitions. This chemical behavior led us to look for a more universal trend in ionization energies. Owing to historical reasons, and also to the measurement process, impurity binding energies are usually referred to one of the host band edges (host referred binding energies [HRBE]). Of course this is the natural choice for the shallow, band-derived state; for deep-level impurities, however, this procedure is not very illuminating. The compilation of a large amount of data for acceptor and donor levels of 3d impurities in the III-V host GaAs, GaP, and InP and on ZnS, ZnSe, CdTe, and other II-VI compounds failed to reveal chemical regularity for individual impurities. If, on the other hand, we look for a unified origin (the "vacuum level") for measuring binding energies, we find a very near constancy in values for the same impurity in compounds within the same class, of the vacuum-referred binding energies (VRBE)[12]. This can be seen in figure 4, where are shown the ionization levels of several impurity systems, when we refer the host band edges to a common vacuum level. This regularity can be traced to the antibonding nature of the impurity states. The interaction of the d states of the ion with the p states of the host anion (around the top of the valence band) produces a bonding resonance in the band (t_2^R) and an antibonding impurity state (t_2^H); the amount by which the antibonding states is pushed up relative to the impurity d state is fairly constant because of the interplay between energy proximity and strength of repulsion of the (higher up) p states. On the other hand, as mentioned in Section 3, multiplet effects are very similar for a given impurity in compounds of the same class; hence, VRBE's are also very similar. With this new choice of origin, we are able to understand why a given impurity introduces a deep acceptor in such a compound, but not in another (e.g., Mn in GaP and GaAs), and also to predict the localization ionization levels where experimental information is lacking just by reference to a known impurity system.

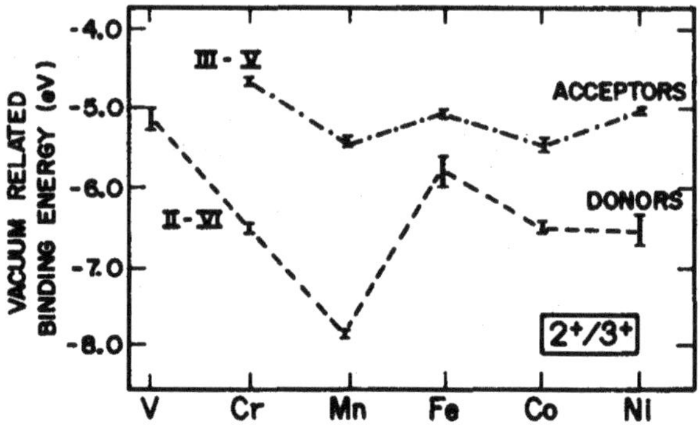

Figure 4. Binding energies of substitutional 3d impurities in semiconductors referred to the approximate host crystal vacuum level. The vertical bars denote the range of ionization levels for a given impurity within a class of compounds. The lower curve depicts donors in the II-VI hosts ZnS, ZnSe, ZnTe, CdSe, and CdTe, whereas the upper curve shows acceptor levels in the III-V hosts GaAs, GaP, and InP.

As very interesting corollaries, VRBE's may help us to follow the valence band edge energy in puzzling situations, namely (1) mixed alloys and (2) unstrained heterojunctions. In the first case, for instance, Cu impurities in $Ga_{1-x}A\ell_xAs$, with the VRBE of the impurity fixed, the variation with x of the HRBE will lead us to the variation with composition of the valence band edge energy[13].

REFERENCES

1. Clerjaud, B., J. Phys. C 18, 3615 (1985).
2. Hemstreet, L.A., Phys. Rev. B 22, 4590 (1980).
3. Fazzio, A., Caldas, M.J. and Zunger, A., Phys. Rev. B 30, 3430 (1984).
4. Watanabe, S. and Kamimura, H., Materials Science and Engineering B3, 313 (1989).
5. Zigone, M., Galtier, P., Martinez, G. and Deveaud, B., Metallurgical Society of AIME, 1059 (1985); Zigone, M., Buisson, H.R. and

Martinez, G., Materials Science Forum 10-12, 663 (1986).

6. Lima, G.A.R., Sardela, M.R., Fazzio, A. and Mota, R., Solid State Comm. 69, 461 (1989).

7. "Landolt-Börnstein Numerical Data and Functional Relationships in Science and Technology", ed. by O. Madelung, Vol. 17 (Springer, Berlin, 1982).

8. The λ_e and λ_t parameters have been determined upon the condition that the $^1A_1$ state keeps above the ^5E state.

9. Makiuchi, N., Macedo, T.C., Caldas, M.J. and Fazzio, A., Defect and Diffusion Forum 62, 63, 145 (1989).

10. Ushakov, V.V., Gippius, A.A., Dravin, V.A. and Kantser, S.F., Sov. Phys. Semicond. 15, 890 (1981).

11. Griffith, J.S., "The Theory of Transition Metal Ions" (Cambridge University Press, London, 1971) and Fraga, S., Karkowski, J. and Saxena, K.H.S., "Handbook of Atomic Data" (Elsevier, Amsterdam, 1976).

12. Caldas, M.J., Fazzio, A. and Zunger, A., Appl. Phys. Lett. 45, 671 (1984).

13. Zunger, A., Phys. Rev. Lett. 54, 849 (1985).

ON THE SHUBNIKOV-dE HAAS EFFECT IN SEMICONDUCTORS

André B. Henriques

Instituto de Física, Universidade de São Paulo,
05508 São Paulo, Brazil

ABSTRACT

A brief discussion of the physics and applications of the Shubnikov-de Haas effect in semiconductors is presented. Illustrative measurements, performed on a modulation doped AlGaAs 150Å quantum well, allow to determine the carrier concentration and Dingle Temperature.

This lecture will address the physics of the low temperature oscillations of the electrical resistance, which are observed in degenerate semiconductors (the Shubnikov-de Haas (SdH) effect). The study of these oscillations is a powerful method for the characterization of novel semiconductor heterostructures, determination of the scattering processes [1,2], impurity states [3], spatial distribution of electrons in modulation doped (MD) quantum wells (QW) [4], and the applications of the method go beyond semiconductors, as in the contemporary studies of low-dimensional organic conductors [5]. Oscillations in the magnetoresistance were first observed in 1930 [6], but it was not until the late 1950's that a first theoretical treatment [7] obtained an analytical expression for the oscillations in the limit of a large filling factor $F \gg \hbar\omega$, were F is the Fermi energy and $\hbar\omega$ is the cyclotron energy, equal to the separation between Landau levels (LL). A modified Boltzmann equation approach was used, and the result is applicable to the longitudinal SdH effect (when the current is made to flow in the same direction as the applied magnetic field). For the transversal configuration, different approaches come to formulae with the same features, i.e. an oscillating term in 1/B modulated by an approximately exponential factor. Several extensive reviews on the subject can be found in the literature [8], and more recently in relation to two-dimensional (2D) semiconductor structures [9,10].

We will consider here the simplest n-type direct-gap degenerate semiconductor, with isotropic and parabolic conduction band, so that an effective-mass treatment is appropriate. It is essential to account for the varying density of states in the conduction band (a growing function of energy $\sim E^{1/2}$ in zero magnetic field in the 3D material), and this requires a statistical approach. In equilibrium, the electron gas is described by the Fermi-Dirac distribution, which is an even function of the Bloch's wavevector **k**. The velocity of carriers being proportional to **k**, this means that in equilibrium there is an equal number of electrons moving in all directions, so no net current exists. When current is made to flow through the sample, one way to describe this is by the equilibrium distribution function, infinitesimally displaced in **k**-space, and this is the Boltzmann equation approach. The finite conductivity of the semiconductor, and hence a fixed displacement of the distribution function for a given applied electrical field, is obtained by introducing a **k**-dependent momentum relaxation time, which can be calculated from a quantum-mechanical perturbation treatment on considering specific scattering mechanisms. It follows from this picture that the electrical response of the semiconductor to an applied voltage is determined by the carriers in an energy range $\sim kT$ around the Fermi surface, and it is the efficiency of the scattering at these energies which determines the conductivity. At low temperatures, when the electrons thermalize to the bottom of the conduction band, the phonon and impurity scattering at the Fermi energy is predominantly elastic, given that $F \gg kT$ (note that kT is the average energy of the acoustic phonon which can be involved in the scattering act). On the other hand, the rate of scattering is proportional to the final density of states, and any change in the density of states at the Fermi energy will be followed by a change in conductivity.

When a strong magnetic field is applied, the movement of the electrons normal to the field is restricted to localized cyclotron orbits (perpetual in the absence of scattering), and the transversal velocity is no longer a good quantum number. However, the electron longitudinal velocity operator commutes with the Hamiltonian, so that in order to describe the longitudinal conductivity a Boltzmann equation approach is still possible. The density of states now presents singularities at the LL energies $E_N=(N+\frac{1}{2})\hbar\omega$, and the number of electrons that can be placed in one given LL is a growing function of the intensity of the magnetic field. When the intensity of B is increased from zero, LL's of decreasing indexes cross the Fermi energy; under these conditions, the scattering rate at the Fermi energy is a maximum, and the longitudinal conductivity σ_{zz} is a minimum. In contrast, the scattering process is the origin of the transversal conductivity σ_{xx}, so that the transversal conductivity will be a maximum at the fields when $F=(N+\frac{1}{2})\hbar\omega$. However, even for an isotropic material, in a magnetic field the conductivity and resistivity become anti-symmetric tensor quantities, and the relation between them is given by

$$\rho_{xx} \cong \sigma_{xx}/\sigma_{xy}^2 , \quad \rho_{zz} = 1/\sigma_{zz} \quad .$$

The approximate sign is valid in sufficiently strong magnetic fields, when $\sigma_{xy} \gg \sigma_{xx}$. Therefore the SdH effect will be characterized by minima both in the longitudinal and transversal resistivities, which occur at the condition $F=(N+\frac{1}{2})\hbar\omega$.

It is then clear that the experimental conditions necessary for the observation of the SdH oscillations are as following:

1) A degenerate semiconductor, so that the Fermi energy lies above the bottom of the conduction band ($F>E_c$);

2) The extent of the energy region, which determines the conductivity, is less than the separation between LL's ($kT<\hbar\omega$);

3) The carriers perceive an oscillating density of states, and this requires that the energy uncertainty of the Bloch wavepacket is much less than the separation between LL's. This condition can be expressed in terms of the electron mobility μ: $\mu B>>1$ (MKS). The higher the mobility of the semiconductor, the lower the field requirements for the experimental observation of SdH effect.

Now consider an ideal 2D semiconductor. In a magnetic field, the density of states is given by (in the abscence of spin-splitting)

$$D(E) = 2eB/h \sum_{N\geq 1} \delta[E-(N+\tfrac{1}{2})\hbar\omega]$$

For a sheet carrier density n_s, at T=0K an integer number N of LL's will be filled for a field of strength B_N

$$n_s = N\ 2eB_N/h \qquad (1)$$

and the transversal conductivity will be a minimum according to the Pauli exclusion principle. The quantum limit is attained when the filling factor is equal to 1. The intensity of the fundamental field is then

$$B_F = hn_s/(2e)$$

In terms of B_F, equation (1) can be rewritten as

$$N=B_F/B_N$$

Hence consecutive minima in the resistivity oscillations at fields B_N can be associated with consecutive integers N, and a plot of these integers against $1/B_N$ will produce a straight line of slope B_F, which crosses the origin. The slope of this line determines the carrier density.

An alternative way to describe the SdH oscillations: from equation (1), the minima in resistivity occur for fields of intensities given by

$$B_N^{-1} = 2eN/(hn_s) \quad ,$$

and a plot of resistivity against 1/B will show a periodicity of $2e/(hn_s)$.

In practice, quasi-2D degenerate semiconductors can be obtained with $Al_xGa_{1-x}As/GaAs$ heterostructures and modulation doped quantum wells (MDQW) by use of modern growth techniques [11]. For an AlGaAs MDQW, a thin layer of the lower bandgap GaAs is grown between thick layers of the higher bandgap AlGaAs, and the unidimensional QW formed, of depth equal to the conduction-band offset, constitutes a trap for electrons. For refined growth techniques, the crystallinity of the structure approaches perfectness, and as a result of the spatial separation between carriers and ionized donors, electrons can attain mobilities far exceeding the Brooks-Herring prediction for bulk materials [12].

In relation to the SdH effect, the 2D behaviour will arise when the width of the confining layer becomes less than the diameter of the cyclotron orbit. The harmonic oscillator wavefunctions, which describe the LL's, are multiplied by an exponential factor, which determine the cyclotron orbit to be of radius$(h/eB)^{\frac{1}{2}}$. Under these conditions, the period of the oscillations will be determined solely by the normal component of the magnetic field, and a simple rotation of the sample in the magnetic field is a characterization technique which can be used to identify the presence of 2D electron gas in a heterostructure [see for example 13]. As the intensity of the magnetic field increases, the radius of the cyclotron orbit is reduced, and for wide MDQW the 2D behaviour changes to isotropic, as shown by Yoshino et al [14].

In order to illustrate this lecture, measurements were taken on a 150Å AlGaAs MDQW at 2K. This is sample G258 in reference [4], where further details of the structure can be found. In this case, the sample was submitted to full illumination with light of 1.65eV at 2K, leading to an increase in the carrier concentration as a result of the persistent photoconductivity (PPC) effect. Figure 1. shows the SdH oscillations in the transverse configuration (the insert shows the inverse field apodised Fourier transform, from which the fundamental field is found to be 20.8T; it should be noted that Fourier analysis is a particularly useful method when more than one subband is occupied, as in some cases it allows the determination of the the population of each of the subbands [10,15]).

Figure 2 shows the filling factor as a function of inverse field, and the slope of the curve is B_F=20.8T, in agreement with the Fourier analysis. From the value of B_F, the sheet carrier density is $n_s=1.0\times10^{12}cm^{-2}$, which is significantly larger than found in [4] ($n_s=7\times10^{11}cm^{-2}$ for only partial illumination of the sample). Further, an

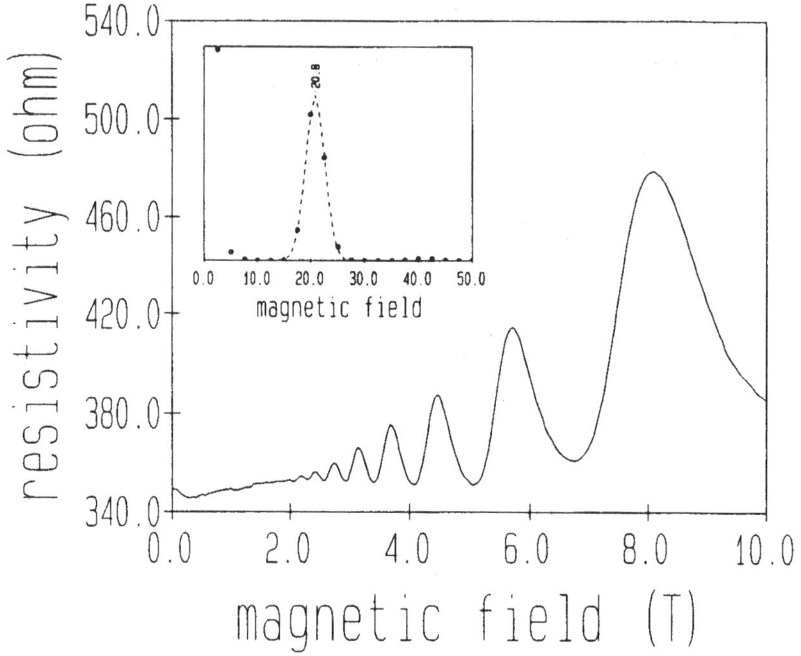

Figure 1. Transversal SdH effect on a 150Å AlGaAs MDQW. The insert shows the inverse field apodised Fourier Transform. The dashed line shows a Gaussian fit to the Fourier Transform, from which the fundamental field is found to be 20.8T.

attempt was made to fit the lower field N=10,11 (the interval 1.7-2.0T) oscillations with the theoretical expression for the resistivity oscillations [16], as derived from the conductivity result [17], in the limit of a large filling factor:

$$\Delta \rho_{xx} = 2\, \rho_{xx}\, \beta \exp(-\beta_D)\, \cos(2\pi B_F/B)/\sinh(\beta) \qquad (2)$$

where

$$\beta_{(D)} = 2\pi^2 k T_{(D)}/\hbar\omega \quad ,$$

T is the electron temperature and T_D is the Dingle temperature, which is characterized by the lifetime of the particle states τ_E

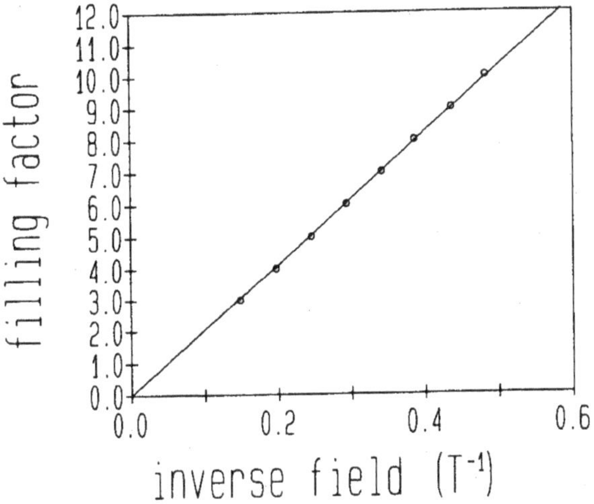

Figure 2. Filling factor against inverse fields at which minima in resistivity are observed in Fig.1. The slope of the fitting solid line is B_F=20.8T, in agreement with Fig.1.

$$T_D = \hbar/(2\pi k\tau_E)$$

In order to exclude the monotonously rising background, two approaches were used: filtering the Fourier Transform (a removal of the constant and lowest frequency harmonic amplitudes with a subsequent inverse Fourier transformation) and a subtraction of a seemingly appropriate background. Both methods come to the same result, but it was found that the removal of even the lowest harmonic lead to a subtle

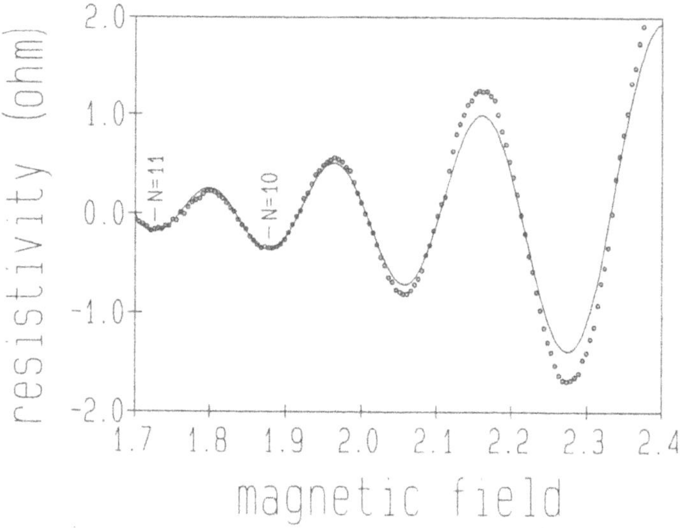

Figure 3. Fit (solid line) of the low field (1.7-2.0T) resistivity oscillations (circles) at 2K with expression (2). From the fit T_D=13.9K.

deformation of the spectrum, and the second method was preferred. The input power per electron in these measurements was $1x10^{-17}W$, and in this case the electrons are in thermal equilibrium with the crystal lattice [1], allowing the fit to be obtained with a single adjustable parameter - T_D (an electron effective mass of 0.068 was used). As shown in Fig.3, the 1.7-2.0T fit deviates from the experimental data as the filling factor is decreased, as expected from the breakdown of the theoretical assumptions. The best fit gives T_D=13.9K, corresponding to a quantum lifetime of τ_E=0.1ps. Using the result of [4], the value of τ_E found is approximately half the magnitude of the scattering rate τ_k which can be deduced from the mobility $\mu=e\tau_k/m^*$ (μ=8.88MKS for this sample [4]). This is a consequence of the fact that the quantum lifetime is determined by the overall scattering, whereas τ_k is insensitive to low angle scattering.

ACKNOWLEDGEMENTS
This work was sponsored by the Programa BID/USP, Projeto No.21.11(070). Experiments were performed at the Clarendon Laboratory, Oxford, UK, and the author is indebted to Dr R.J.Nicholas for permission to use the high-field facilities and loan of the AlGaAs sample, to Dr D.R.Leadley for valuable discussions, to Dr F.L.Pratt for loan of laboratory equipment.

REFERENCES

1. H.Sakaki, K.Hirakawa, J.Yoshino, S.P.Svensson, Y.Sekiguchi, T.Hotta and S.Nishii, Surf. Sci. 142, 306-313 (1984)

2. D.R.Leadley and R.J.Nicholas. to appear in Semicond.Sci.Technol.

3. G.Gornik, Proceedings of the 1989 IV Brazilian School of Semiconductor Physics (to appear)

4. J.J.Harris, J.M.Langemaat, S.J.Battersby, C.M.Hellon, C.T.Foxon and D.E.Lacklison, Semicond.Sci.Technol. 3, 773-780 (1988)

5. F.L.Pratt, A.J.Fisher, W.Hayes, J.Singleton, S.Spermon, M.Kurmoo and P.Day, Phys.Rev.Lett.61, 2721 (1988)

6. L.Shubnikov and W.J.de Haas, K.Akad.Amsterdam, Proc. 33, 4, 363 (1930)

7. P.N.Argyres, J.Phys.Chem.Solids **4**, 19-26 (1959)

8. R.L.Peterson, 'The Magnetophonon Effect', Semiconductors and Semimetals **10**, 221 (1975); I.M.Tsidilkovskii, 'Electrons and Holes in Semiconductors', Nauka Press, Moscow 1978, p.255-283

9. T.Ando, A.B.Fowler and F.Stern, Rev.Mod.Phys. 54, 437 (1982)

10. R.J.Nicholas, 'High field Magnetotransport', *in* 'Physics and Applications of Quantum Wells and Superlattices', NATO ASI Series **170**, 217 (1987)

11. K.Ploog, 'Molecular Beam Epitaxy of Artificially layered III-V Semiconductors on an Atomic Scale' *in* 'Physics and Applications of Quantum Wells and Superlattices', NATO ASI Series **170**, 217 (1987)

12.G.Weimann and W.Schlapp, 'Transport Properties of Two-Dimensional Electron and Hole Gases in GaAs/AlGaAs heterojunctions', Springer-Verlag Solid State Sciences Series 67, 33-41, 1986

13. S.K.Haywood, A.B.Henriques, D.F.Howell, N.J.Mason, R.J.Nicholas and P.J.Walker, Institute of Physics Conference Series 91, 271 (1988)

14. J.Yoshino, H.Sakaki, T.Hotta, Surf. Sci. 142, 326-331 (1984)

15. W.Zhao, F.Koch, J.Ziegler and H.Mayer, Phys.Rev.B 31,2416 (1985)

16. P.T.Coleridge, R.T.Stormer and R.Fletcher, Phys.Rev.B 39,1120 (1989)

17. T.Ando, J.Phys.Soc.Japan 37, 1233 (1974)

ELECTRONIC STRUCTURE OF POLYMERS

Celso P. de Melo

Departamento de Física
Universidade Federal de Pernambuco
50739 Recife PE
BRAZIL

ABSTRACT

This course presents a short review of how techniques borrowed from condensed matter physics can be applied to the calculation of the electronic structure of polymers. Emphasis is given to the study of the electronic and transport properties of conducting polymers: it is shown that the use of transfer functions and renormalization ideas can lead to simple analytical solutions to the problem of how conformational defects affect the electronic structure of conjugated polymers.

1. INTRODUCTION

The electronic structure of polymers is an area of study where condensed matter and molecular physics meet. The complexity of the electronic interactions in these molecules is intrinsecally associated to their extended nature (a polymer can be composed of thousands of repeated elementary unities) and amorphous character (for polymers crystallinity is the exception rather than the rule) which leads to non-negligible couplings among chains. Accordingly, to describe the macroscopic behavior of polymers one is required to use techniques developed for condensed matter physics problems. On the other hand, local interactions may play an important role on certain aspects of the optical, electronic and transport properties of polymers; to describe these, a molecular physics (or quantum chemistry) approach can be more convenient.

Within the scope of each different point of view useful

insights about the behavior of real systems can be acquired through the use of convenient models developed to treat specific problems. As usual in such situations it is important to keep in mind the underlying assumptions of the adopted model and try not to stretch it beyond the appropriate limits. Different problems usually require different models and, hence, the complete description of the observed optical, electronic and transport properties of polymers has to resort to a mosaic of distinct treatments.

This need for alternating point of views puts an extra burden on the student interested in the electronic structure of polymers: to be fluent on a variety of approaches which range from standard quantum chemistry techniques to others based upon advanced condensed matter concepts. In this set of lectures I will try to present a self-contained description of the basic ideas concerning the physics of conjugated of polymers; emphasis will be given to the discussion of how a combination of a Green's function formalism with the use of transfer functions can be applied to obtain the general features of the electronic structure of polymers in a simple and analytical way. Although most of the treatments presented in this course could easily be applied to the electronic structure of polymers in general, the guideline of the following discussion will be the study of the interesting electronic and transport properties of conducting polymers. We begin by a brief historical sketch of the major developments in the area of conducting polymers.

1.1. Polymers as Conducting Materials

Due to the localized character of the covalent bonds between carbon atoms, up to the late 60's organic compounds were usually thought of as uninteresting materials concerning to electrical properties. However, one can note that already in 1964 Little had suggested that a possible way to find superconductivity at room temperature was to look to special electronic interactions in long conjugated (i.e. with alternating single and double CC bonds) chains substituted by chromophore groups[1] (in the 80's organic superconductors have been found but they are neither polymeric nor follow Little's suggested mechanism; even so, Little's original insight is nowadays considered an important landmark in the field).

Films of the trans isomer (Fig. 1) of polyacetylene (PA) were obtained for the first time by Shirakawa[2] in 1974. The films were of fibrilar and amorphous nature and electrical and optical characterization indicated the polymer to be a semiconductor with an estimated gap of 1.4 eV. Surprisingly, it was soon recognized that upon n- or p-doping (i.e. exposure of the material to reducing or oxydizing agents) samples of trans-PA would experiment a dramatic change on their electrical conductivity[3]. Typically, the conductivity of the material would increase by at least 13 orders of magnitude for concentration of

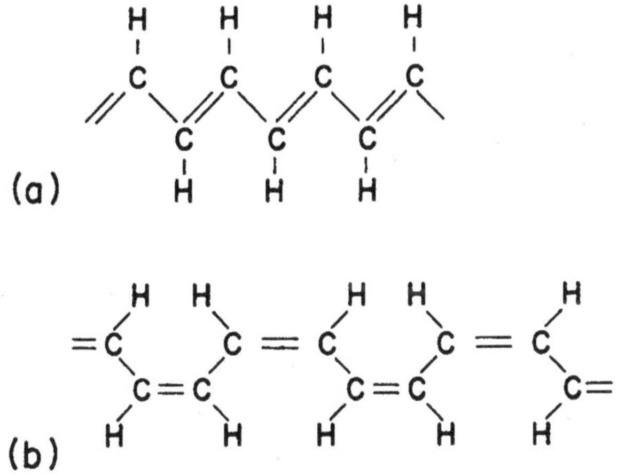

Fig.1. a) Trans- and b) cis- isomers of polyacetylene.

3-5% of dopants[4,5,6], as schematically represented in Fig.2.

In fact the synthesis of trans-PA by Shirakawa settled on itself a long standing question about the true nature of the ground state of the material[7,8]. Simple chemical intuition would suggest a benzene-like resonance along the conjugated chain and consequent homogeneous CC bondlengths. The corresponding band structure would be typical of a metal since twice as much states would be available than electrons to fill them. But as pointed earlier by Peierls[9], half-

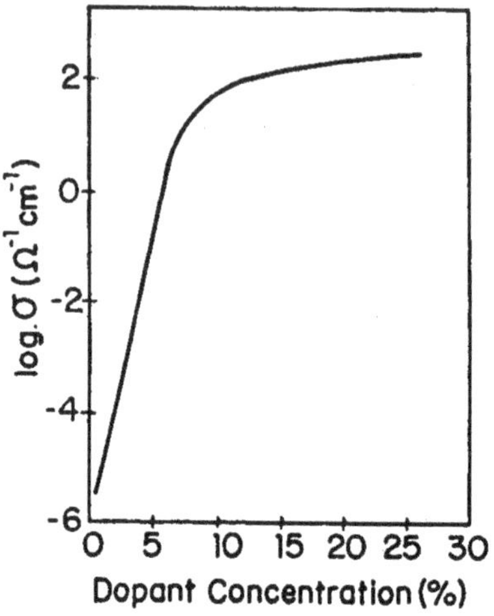

Fig.2. Schematic representation of the change on electrical conductivity of PA as a function of dopant concentration.

filled bands are not energetically favored in one-dimension; the polymeric chain should then "dimerize"[10)] (i.e. to present alternating CC bondlengths - see Fig. 3), and an energy gap is expected to open between occupied and virtual states[1]. Should the Peierls instability be the dominant factor, the stable configuration for the material would correspond to a semiconductor. This is what was finally observed for Shirakawa's films[4,5,6)].

The discovery of the remarkable effect doping had upon the electrical conductivity of trans-PA was a turning point in the field. In a matter of months several different experimental results made clear that new physics concepts had to be invoked to explain the unusual electronic properties of the polymer. For instance, it was recognized that unpaired spins should be present even in pristine samples of both cis- and trans-PA since EPR signal could be observed for both isomers; more surprisingly, for the later the narrow linewidth signal was characteristic of delocalized spins[12)]. At the same time, optical

---

1 - Peierls' instability can be considered as the equivalent of the molecular Jahn-Teller effect[11)] for infinite one-dimensional systems.

absorption spectra indicated that upon doping new states were introduced in the gap[13].

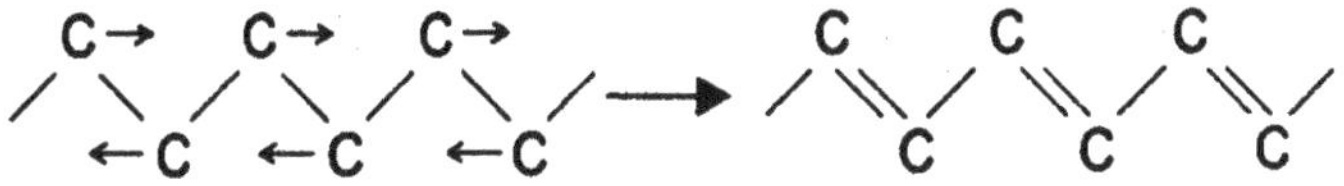

Fig.3. Dimerization in trans-PA occurs to minimize the electronic energy.

The overriding majority of these experimental results was rationalized with the introduction of the so-called soliton model by Su, Schrieffer and Heeger[14] (SSH) in 1979. In this context solitons are understood as conformational defects mobile along the chains. The mobility would result from the special symmetry of trans-PA in which a defect would connect two isoenergetic chains, and from the estimated small value for the potential barrier for chain reorganization in the region of the defect (see Fig. 4). In pristine samples, solitons

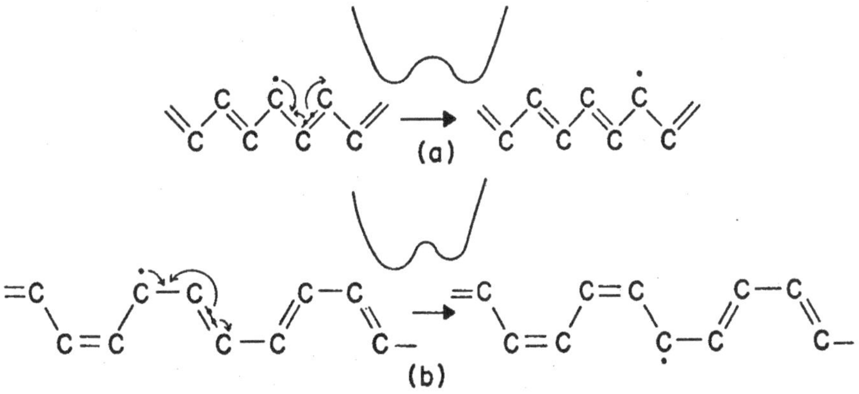

Fig.4. Due to the small elastic constant of the chain, conformational rearrangement is possible in the region of the defect: in trans-PA (a) mobility arises from the degeneracy between initial and final configurations which does not exist in cis-PA (b).

should be neutral but have unpaired spin (in chemical terminology a soliton would correspond to a neutral radical)[15].

Due to the fibrilar nature of the material, doping by oxydizing or reducing agents would consist in the simple exchange of charge between the conjugated chain and the dopant (these should be nested in the large interchain volume): charged solitons, which possess no spin, would be the charge carriers. Although the SSH model glosses over some finer points (as the fact that real chains are necessarily finite and therefore interchain charge transport mechanisms should at least complement soliton transport) it sucessfully explained most of the then known phenomenology of PA[4,5,6].

However, soon it was observed that the so called "chain-bridged" polymers (Fig. 5) present transport properties similar to trans-PA[16]. The corresponding chains were more similar to cis- than to trans-PA and hence conformational defects of the previous kind would connect two energetically inequivalent chains.

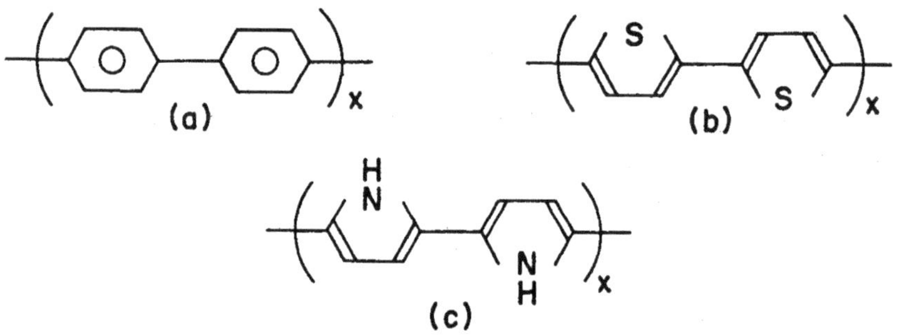

Fig.5. Polyparaphenylene, PPP, (a), Polythiophene, PTP, (b), and polypyrrol, PPY, (c) present transport properties similar to polyacetylene.

The way to allow for mobility here is to consider the possibility of interaction between solitons[17]: the resulting structures would again connect isoenergetic chains and be mobile.

Polarons (Fig. 6) was the name given to this new type of defects which bear charge-spin relationships different from those of solitons (Fig.7).

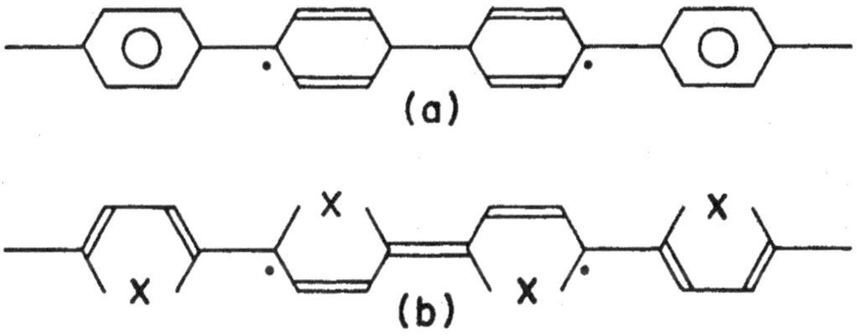

Fig.6. Polarons in a) PPP and b) PTP (X≡S) and PPY (X≡NH) chains.

Once again it seemed that if one could sweep under the rug a few lingering questions (as the nature of interchain charge transport in real systems) polarons could come close to be the generic mechanism for conductivity in conjugated polymers. But, as ever, Nature knows best: in the last few years it was recognized that in polyaniline

	$S^•$	S^+	S^-	P	P^+	P^{++}	P^-	P^{--}
Charge	0	+1	-1	zero	+1	+2	-1	-2
Spin	½	0	0	0	½	0	½	0

Fig.7. Charge-spin relationships are distinct for solitons (a) and polarons (b).

(Fig. 8) conductivity is affected by changes on the pH of the medium[18]. Since pH variations correspond to different amounts of proton doping, in the process the number of electrons is not affected. Therefore only structural changes are left to account for the increase of conductivity and although the true origin of conductivity in polyaniline is still not understood it has been suggested that structural disorder may affect the electronic structure leading to a typical metallic behavior[19]. But by now we have learned that there are few chances that a single mechanism could ever be identified as the universal answer to the conductivity of polymers.

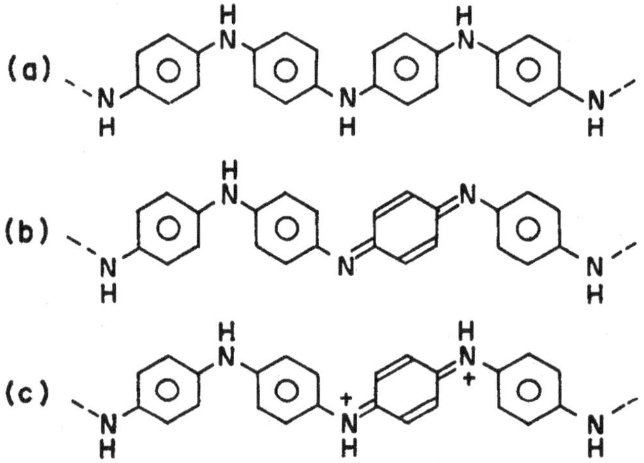

Fig.8. Leucoemeraldine (a), and emeraldine (b) are alternative forms of polyaniline.Protonated emeraldine (c) is believed to be the conducting form of the material under acidic (low pH) conditions.

2. GREEN'S FUNCTION DESCRIPTION OF THE ELECTRONIC STRUCTURE OF POLYMERS

2.1. Introduction

The Huckel Model is a sucessful quantum chemistry first approach to the treatment of the electronic structure of conjugated molecules[20]. The method is based upon the assumption of σ-π separability, the idea that since the highest occupied and the lowest unoccupied orbitals of conjugated systems are of π character the role of the σ electrons can be approximated to one of providing an effective field in which the π electrons move[20]. We are then left with one orbital per atom, and the overlap matrix is taken approximately equal to the identity. The hamiltonian matrix is simplified by admiting only diagonal and first neighbors interactions, which are taken as free parameters (α and V , respectively). Diagonalization of the hamiltonian matrix would provide the desired eigenvectors (molecular orbitals) and eigenvalues (orbital energies); the parameters α and V could then be adjusted to best reproduce experimental data. In view of the crudeness of the model, it is surprising to learn how successful is the Huckel approximation in giving the most important qualitative features of the electronic structure of conjugated molecules[21]. In fact, the Huckel method was the most important tool for theoretical investigation of the properties of organic molecules up to the late 60's when the age of computers allowed the introduction of more sophisticated quantum chemistry treatments[22].

For conjugated polymers the σ-π approximation still holds as a good starting point: for instance, in trans-PA the total bandwidth is of the order of 10 eV, roughly the same order of the σ-σ* energy gap. Also, physicists have developed similar ideas to the Huckel model to treat one-dimensional (1-D) systems[4]. The overlap approximation is formally equivalent to use of localized Wannier functions[23] as the appropriate basis set. Tight-binding Approximation[24] with first-neighbors interaction would then be the physicists' parlance to the same set of assumptions as those of a molecular Huckel treatment.

For infinite systems direct diagonalization of the hamiltonian matrix is out of question. One of the most convenient ways of avoiding the problem is to introduce the resolvent or Green's operator[25)]

$$\hat{G}\ (E) = \frac{1}{E - \hat{H}} \tag{1}$$

In an orthonormal basis $|i\rangle$, the overlap matrix is diagonal, $\langle i|j\rangle=\delta_{ij}$, and the Green's operator can be written as

$$E\ G_{ij} = \delta_{ij} + \sum_k H_{ik}\ G_{kj} \tag{2}$$

If we assume the matrix elements of the hamiltonian to be known, the physics is contained in the elements G_{ij}. However, even for a first-neighbors treatment expression (2) represents an infinite set of coupled algebraic equations. Fortunately, things are not so complicated as they look because for periodic systems there are ways to decouple the equations.

Let's consider some illustrative examples of the use of Green's functions for the study of the electronic structure of polymers.

2.1. Homoatomic Regular Chain

Here, by regular we mean a system with homogeneous first-neighbors bondlengths, as that of Fig. 9 (this could be a model for polyacetylene if Peierls instability did not occur).

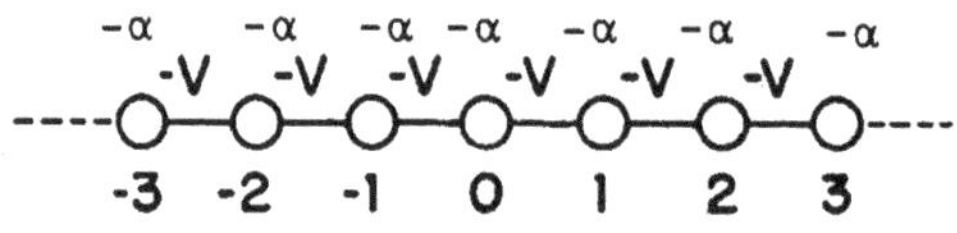

Fig.9. Homoatomic regular chain.

The corresponding hamiltonian matrix elements can be written as

$$H_{ij} = \begin{cases} -\alpha\,, & \text{for } i=j \\ -V\,, & \text{for } i=j\pm1 \\ \text{zero}, & \text{otherwise} \end{cases}$$

For reasons that will be clear shortly let's look first to the diagonal elements $G_{i,i}$. If we now use expression (2) for i=j=0, we obtain the infinite set of coupled algebraic equations

$$(E + \alpha)G_{0,0} = 1 - V\,G_{-1,0} - V\,G_{1,0} \tag{3.1}$$

$$(E + \alpha)G_{1,0} = -V\,G_{0,0} - V\,G_{2,0} \tag{3.2}$$

$$(E + \alpha)G_{2,0} = -V\,G_{1,0} - V\,G_{3,0} \tag{3.3}$$

$$\ldots \qquad \ldots \qquad \ldots\ldots$$

One can note that all but the first of the above equations can be written generically as

$$(E + \alpha)G_{n+1,0} = -V\,G_{n,0} - V\,G_{n+2,0} \quad . \tag{4}$$

If we define the transfer function[26] T(E) as

$$T(E) = \frac{G_{n+1,0}}{G_{n,0}} \quad \text{for } n = 0,1,2,\ldots \tag{5}$$

and substitute in Eq.(4) we obtain directly[2]

$$T(E) = \frac{-(E + \alpha) \pm \sqrt{(E + \alpha)^2 - 4\,V^2}}{2\,V} \tag{6}$$

The lowest energy **states** should be symmetrical about site 0;

---

2 - In fact expression (5) need not be introduced as an ansatz: it can be derived from Euler's theorem for continuous fractions[27]. I thank F. Souza for bringing this point to my attention.

hence $G_{-1,0}=G_{1,0}=TG_{0,0}$ and Eq.(3.1) gives us finally

$$G_{0,0}(E) = \pm \frac{1}{\sqrt{(E+\alpha)^2 - 4V^2}} \tag{7}$$

Since this system is translationally invariant all off-diagonal elements can be determined as $G_{m,n}=G_{|m-n|,0}=T^{|m-n|}G_{0,0}$.

The physical meaning of the diagonal elements of G is made more transparent if we note from Eq.(7) that for the energy range

$$-\alpha - 2V < E < -\alpha + 2V \tag{8}$$

the function $G_{0,0}$ is purely imaginary. But if we use the identity[28)]

$$\frac{1}{x - x_0} = P \int \frac{\delta(x - x_0)dx}{x - x_0} - i\pi\delta(x - x_0) \tag{9}$$

where P stands for the Principal Part of the integral, one can see that the imaginary part of G(E) given by Eq. (1) counts the number of states which have energy E. Physicists usually define the local density of states[29)] (lds)

$$\rho_{i,i}(E) = -\frac{1}{\pi} \operatorname{Im} G_{i,i}(E) \tag{10}$$

as a convenient way to find the number of occupied states per energy interval[3]. The continuous energy range for which G(E) is purely imaginary defines the band region: therefore the corresponding states are delocalized in nature.

For chemists a more familiar quantity can be obtained after integration of $\rho_{i,i}(E)$ over the occupied part of the band up to the

---

3 - The local density of states can be probed experimentally through such techniques as photoemission spectroscopy, where photons of known energy and momentum are used to scatter bound electrons[30)].

Fermi energy E_f (i.e. the energy associated with the highest occupied states):

$$n_i = R_{ii} = \int_{-\infty}^{E_f} \rho_{ii}(E)\, dE \qquad (11a)$$

This is clearly the electronic charge (per spin) at site i; similarly integration of the off-diagonal elements would give

$$R_{ij} = \int_{-\infty}^{E_f} \rho_{ij}(E)\, dE\ ; \qquad (11b)$$

R_{ij} is the (one-electron) density matrix related to the familiar charge and bond-order matrix P_{ij} of quantum chemistry[22] by $P_{ij}=2R_{ij}$.

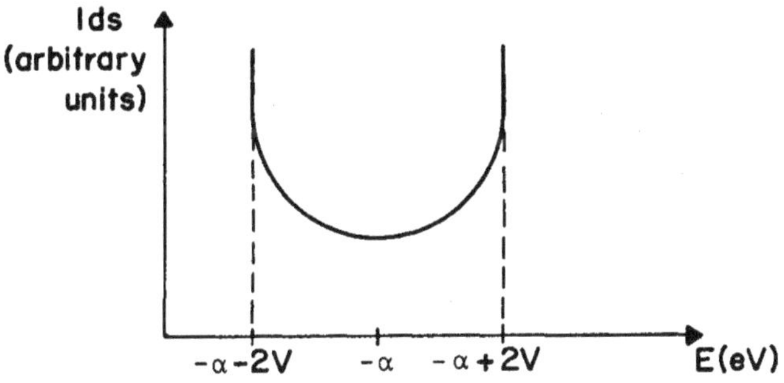

Fig.10. Local density of states for regular homoatomic chain. Since there is no charge transfer in this system, all sites have the same lds.

The lds for the homoatomic chain is represented in Fig. 10. The choice of sign on Eq. (7) is such as to lead to a positive density of states as it should. As we have seen, the homoatomic chain is a model for regular PA: for this system we then obtain a single half-filled band and the predicted behavior is metallic as antecipated.

2.2. Alternant Homoatomic Chain

As we have seen above, the model of Fig. (11) is a more realistic representation of trans-PA as for this polymer the elastic constant of the chain is small enough to make dimerization possible.

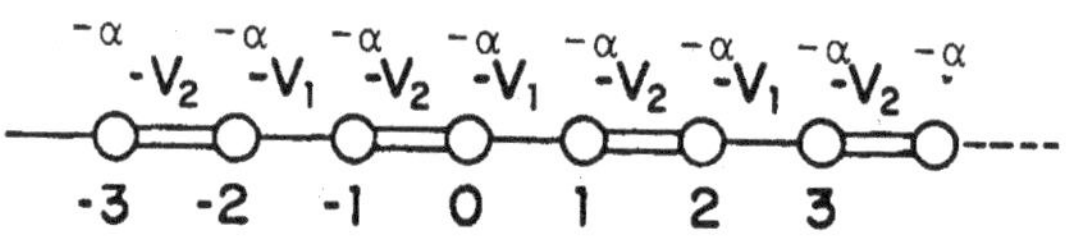

Fig.11. Alternant homoatomic chain.

The hamiltonian matrix elements can be parametrized as

$$H_{ij} \quad \begin{array}{ll} -\alpha & \text{for } i=j \\ -V_1 & \text{for } j=i+1, \text{ even } i \\ -V_2 & \text{for } j=i+1, \text{ odd } i \\ 0 & \text{otherwise} \end{array}$$

Use of expression (2) for site 0 gives directly

$$\begin{array}{l} (E+\alpha)\, G_{0,0} = 1 - V_2\, G_{-1,0} - V_1\, G_{1,0} \\ (E+\alpha)\, G_{1,0} = - V_1\, G_{0,0} - V_2\, G_{2,0} \\ (E+\alpha)\, G_{2,0} = - V_2\, G_{1,0} - V_1\, G_{3,0} \\ \quad \cdots\cdots \qquad \cdots\cdots\cdots \\ \quad \cdots\cdots \qquad \cdots\cdots\cdots \end{array} \tag{12}$$

or for n=0,1,2,...

$$\begin{array}{l} (E+\alpha)\, G_{2n+1,0} = - V_1\, G_{2n,0} - V_2\, G_{2n+2,0} \\ \\ (E+\alpha)\, G_{2n+2,0} = - V_2\, G_{2n+1,0} - V_2\, G_{2n+3,0} \end{array} \tag{13}$$

The alternant chain is still a periodic system and the problem has exact solution if we define the transfer functions[31)]

$$T_1 = \frac{G_{2n+1,0}}{G_{2n,0}} \quad \text{and} \quad T_2 = \frac{G_{2n+2,0}}{G_{2n+1,0}} \tag{14}$$

Substitution of (14) in (13) produces the expressions for T_1 and T_2 as

$$T_1 = \frac{-[\epsilon^2 - (\beta^2 - 1)] \pm \sqrt{[\epsilon^2 - (\beta + 1)^2]\,[\epsilon^2 - (\beta - 1)^2]}}{2\,\epsilon}$$

$$T_2 = -\frac{\beta}{\epsilon + T_1} \tag{15}$$

where we have introduced the notation $\epsilon = (E + \alpha)/V_1$ and $\beta = V_2/V_1$. For the band region we choose the sign on (15) such that a positive lds results. Since T_1 and T_2 are associated to single and double bonds, respectively, we can write $G_{-1,0}=T_2G_{0,0}$ and $G_{1,0}=T_1G_{0,0}$ (note that this problem has no inversion symmetry).

Hence the first of Eqs. (12) gives us finally

$$G_{0,0} = \frac{\epsilon}{\sqrt{[\epsilon^2 - (\beta + 1)^2]\,[\epsilon^2 - (\beta - 1)^2]}} \tag{16}$$

and the corresponding density of states is (Fig. 12)

$$\rho_{0,0}(E) = -\frac{1}{\pi}\,\mathrm{Im}\,\frac{|\epsilon|}{\sqrt{[(\beta + 1)^2 - \epsilon^2]\,[\epsilon^2 - (\beta - 1)^2]}} \tag{17}$$

The first thing to note is that there are two band regions

$$-(\beta + 1) < \epsilon < -(\beta - 1) \quad \text{and} \quad (\beta - 1) < \epsilon < (\beta + 1)$$

For PA there are twice as much states as electrons to fill them. Hence, our result indicates that for this system an energy gap exists between the valence and conduction states, i.e. the material should be a semiconductor. From the experimental values for the total π bandwidth $W=2(\beta+1)V_1 \sim 10,0$ eV and for the energy gap $\Delta=2(\beta-1)V_1 \sim 1.4$ eV, we can estimate β to be of the order of 1.32. It is easy to show that in the limit $\beta \to 1$ we recover the regular chain results.

We have seen that the most interesting physics of polyacetylene comes from the existence of conformational defects as that of Fig.(4). This is the problem we are going to handle now.

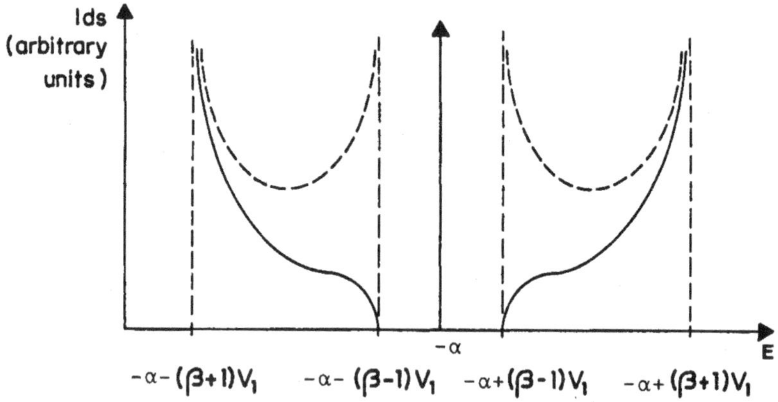

Fig.12. lds for defect-free (dashed line) and single-site defect (full line) in alternant PA.

2.3. Single-Site Defect in an Alternant Homoatomic Chain

With the same hamiltonian parameters as before, expression (2) gives for the model of Fig. 13.

$$
\begin{aligned}
(E+\alpha)G_{0,0} &= 1 - 2\,V_1\,G_{10} \\
(E+\alpha)G_{10} &= -\,V_1\,G_{0,0} - V_2\,G_{20} \\
(E+\alpha)G_{20} &= -\,V_2\,G_{10} - V_1\,G_{30} \\
\cdots\cdots & \quad \cdots\cdots
\end{aligned}
\tag{18}
$$

where we have used the fact that now $G_{-10}=G_{10}$.

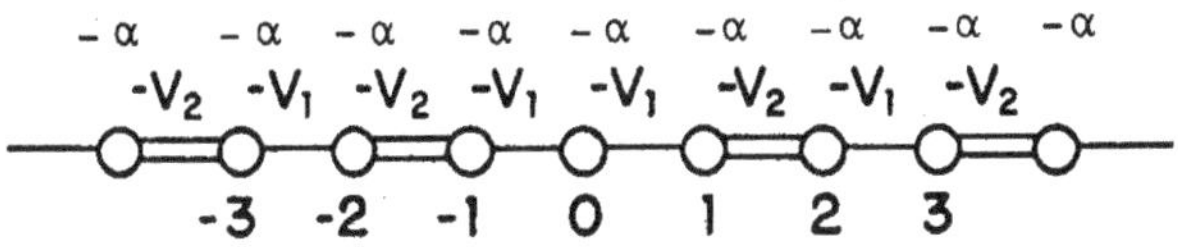

Fig.13. Single-site defect in an alternant homoatomic chain.

We see from the above equations that the periodic part of this problem is identical to the case of the defect-free alternant chain[31). Hence the T functions are given again by (15) and we can write

$$(E + \alpha + 2\, V_1 T_1)\, G_{0,0} = 1 \tag{19}$$

Expressing once more our results in units of V_1 we obtain

$$G_{0,0} = \frac{\epsilon}{(\beta^2 - 1) \pm \sqrt{[\epsilon^2 - (\beta + 1)^2]\,[\epsilon^2 - (\beta - 1)^2]}} \tag{20}$$

We see that although the band region remains the same as before (as it should, since we have a single short-range defect) the lds for the site of the defect

$$\rho_{0,0} = \frac{1}{\pi}\, \frac{\epsilon \sqrt{[(\beta + 1)^2 - \epsilon^2]\,[\epsilon^2 - (\beta - 1)^2]}}{\epsilon^4 + 2\epsilon^2\,(\beta^2 + 1)} \tag{21}$$

is substantially modified (Fig. 12).

The depletion of the valence and conduction band densities of states can be understood if we look at the behavior of $G_{0,0}$ in the region outside the bands. If we write $G_{0,0}=\epsilon/D(\epsilon)$ it is easy to verify that there is a pole for $G_{0,0}$ at $\epsilon=0$: since

$$\lim_{\epsilon \to 0} D(\epsilon) = \epsilon^2\, \frac{(1 + \beta^2)}{(\beta^2 - 1)} \tag{22}$$

the corresponding residue is $(\beta^2 - 1)/(\beta^2 + 1)$. This is exactly the integral of the difference between the two curves of Fig. 12 and therefore no charge rearrangements along the chain must occur as consequence of the breaking on the pattern of bond alternation.

An important result comes if we express the Green operator in terms of the eigenstates of the hamiltonian[32)]

$$\hat{G}(E) = \sum_n \frac{|n\rangle\ \langle n|}{E - E_n} \tag{23}$$

and use the fact that the T function has no common poles with $G_{0,0}$; it is then easy to show that for any element G_{i0} one can write

$$\text{Res } G_{i0}\ (\epsilon = 0) = \psi_i^*\ (\epsilon = 0)\ \psi_0\ (\epsilon = 0) \tag{24}$$

where ψ_i is the coefficient on site i of the expansion of the eigenstate $|n\rangle$ over sites. Hence, we get finally

$$\begin{aligned} &\text{Res } G_{0,0}(\epsilon = 0) = \psi_0^*\ (\epsilon = 0)\ \psi_0\ (\epsilon = 0) \\ &\psi_{-2n-1}(\epsilon = 0) = \psi_{2n+1}(\epsilon = 0) = (T_1T_2)^n T_1\ \psi_o \Big|_{\epsilon = 0} \\ &\psi_{-2n}(\epsilon = 0) = \psi_{2n}(\epsilon = 0) = (T_1T_2)^n \psi_o \Big|_{\epsilon = 0} \end{aligned} \tag{25}$$

We see therefore that the decay of the localized wavefunction associated with the pole $\epsilon=0$ is determined by powers of T_1 and T_2. Hence, outside the band region we must choose the sign on (15) such as to have $|T_1T_2| < 1$, and finally, one find $T_2(\epsilon=0) = \epsilon/(\beta^2-1) \rightarrow 0$, $T_2(\epsilon=0) = -(\beta^2-1)/(\beta\epsilon)$, and $T_1T_2 = -(1/\beta)$.

Using the above results we can map the wavefunction associated to the localized state as

$$\psi_n(\epsilon = 0) = \beta^{-|n/2|} \cos\frac{\pi n}{2}\, \psi_0(\epsilon = 0) \tag{26}$$

As we see from the representation on Fig. 14, the wavefunction associated to the conformational defect has a damped oscillating character: after the central peak at site 0 it decays at the even sites while vanishing at the intermediate ones. Since there is no charge transfer, the localized state represents a damped spin density wave which can be observed in EPR experiments.

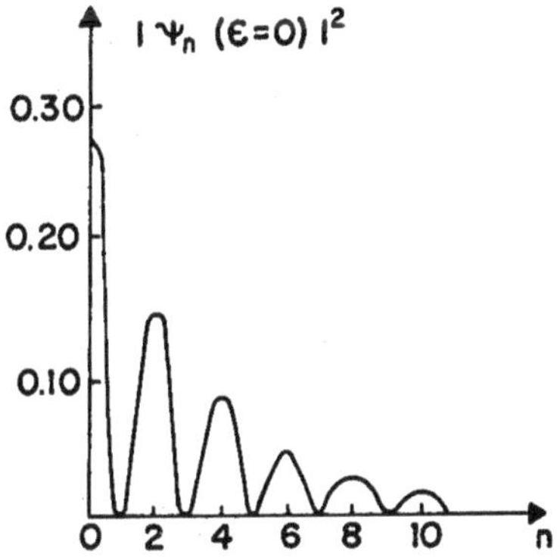

Fig.14. Soliton wavefunction for single-site defect on trans-PA.

2.4. Possible Extensions of the Single-Site Tight-Binding Model

With the simple kind of models we have worked so far we have obtained some important results for the understanding of the electronic structure of polyacetylene: for instance we have verified that the existence of an energy gap between valence and conduction states is a consequence of the bond alternation along the chain and that the breaking of the bond alternation pattern gives origin to a localized spin density wave of oscillating character.

There are however several ways to improve the hamiltonian approximation we have used. For instance, one could be tempted to go beyond the simple Huckel limit and allow for longer range interactions between sites: in fact, a second-neighbors approximation can be handled through the Green's function formalism and lead to interesting results[33], since the standard results for canonical alternant systems (such as the pairing theorem[20]) are not valid in this case. In particular the second-neigbors treatment allows in principle for negative spin densities at the odd sites; however, for the range of parameters relevant for PA the estimated spin densities are rather small and in general the model do not predict any observable differences to the previous result obtained within a first-neighbors approximation.

A more serious criticism of the model used before concerns the choice of a single-site defect to mimic the breaking of bond alternation in PA. If we associate the order parameter $\lambda=-1$ to the pattern of alternation at the far left of the defect, at the far right λ must be equal to +1. For the single-site defect[4] the change on λ should be abrupt, while a more realistic model would assume a defect of a certain width and characteristic tanh form[14] (Fig. 15).

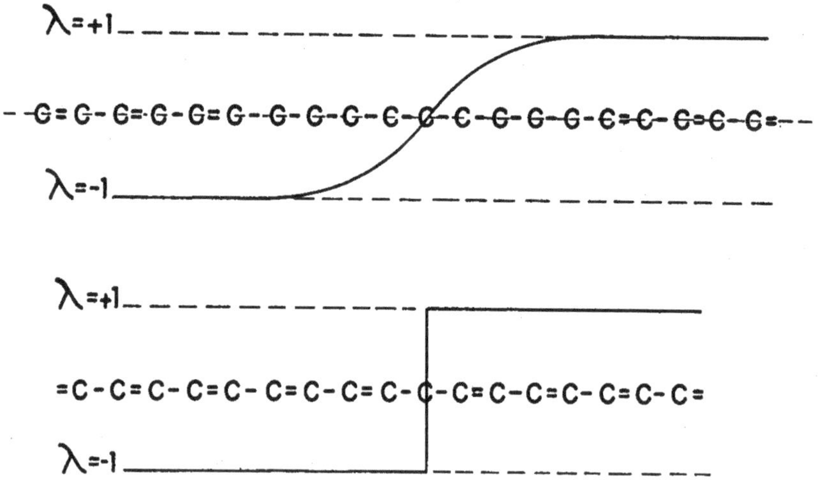

Fig.15. In real systems (a) the conformational defect should involve more sites than admitted in the single site model (b).

There are ways to handle relaxation of defects within the formalism we have used so far. Consider the model of Fig. 16. If we admit that a finite number $2\nu+1$ of sites are affected by the defect (SSH estimated ν to be of the order of 7 in their variational treatment), use of expression (2) from the site $\nu+1$ on will produce

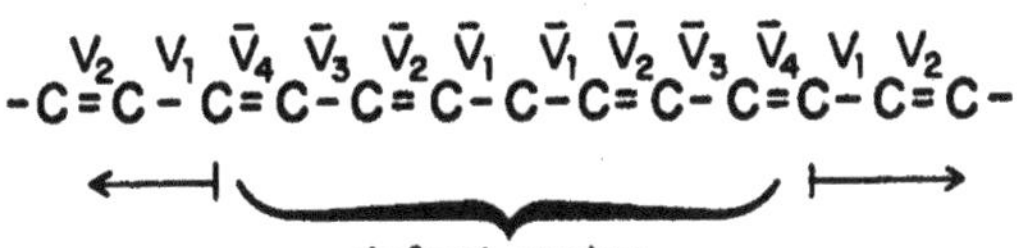

Fig.16. A relaxed defect will have a characteristic width. The periodic part of the problem is identical to that of the defect-free alternant chain.

generic equations of type (13). Hence, the periodic part of the problem is the same as before and we can use the transfer functions T_1 and T_2 of (15). We are then left with a finite cluster of modified sites. If through successive substitutions we eliminate references to the G_{i0} in the cluster region we can obtain the lds for site 0. Although some characteristic oscillations are predicted for the form of the lds[34], the general predictions of the single-site model are vindicated. The importance of this treatment is that the need for successive substitutions brings in the seeds of a powerful technique: the renormalization of the electronic interactions within the Green's function treatment.

3. RENORMALIZATION PROCEDURES FOR THE ELECTRONIC STRUCTURE OF POLYMERS

3.1. Introduction

The systems we have described so far were relatively simple and the success of the treatments we have shown was dependent on our ability to solve the equations for the pertinent transfer functions. It is easy to imagine that for polymers of more complex unit cells (as the chain-bridged polymers of Fig. 5) or long range defects only numerical solutions could be possible.

However, for 1-D systems there is a simple procedure of transforming complex systems into simpler ones through successive substitutions: this is what we will call renormalization technique. As we will show later, this technique will allow the reduction of the electronic structure problem of the chain bridged polymers to the case of PA. To get the feeling of how the renormalization technique works, let's begin by the solution of the diatomic chain problem.

3.2. Renormalization of the Diatomic Chain

A diatomic chain as that of Fig. 17a can be a convenient model to represent polymers as $(SN)_x$ (this is the only known case of a inorganic conducting polymer[35]: in fact, $(SN)_x$ presents a transition to a superconducting phase close to 0 K).

This problem can still be solved through the use of a pair of transfer functions. Let's show however that a simpler way to handle the problem is to redefine the interactions and transform the original chain into a renormalized one which requires only one transfer function to its solution (Fig. 17b).

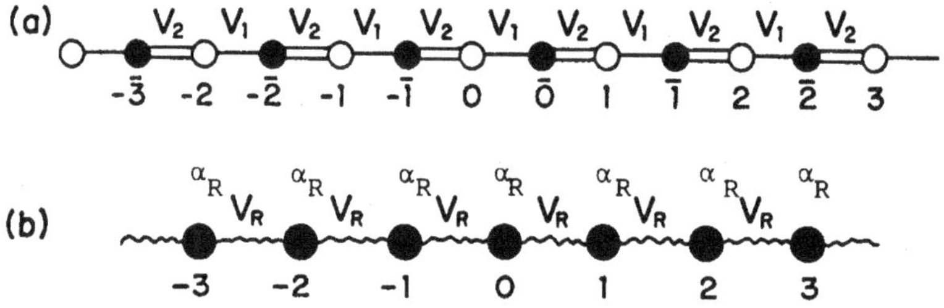

Fig.17. a) Diatomic linear chain and b) its renormalized "homoatomic" equivalent.

In Fig. 17a we have used the overhead bar to indicate the sites which should "disappear" in the renormalization process. Use of expression (2) gives us directly

$$(E + \alpha) G_{00} = 1 - V_1 G_{\overline{0}0} - V_2 G_{-\overline{1}0}$$
$$(E + \overline{\alpha}) G_{\overline{0}0} = - V_1 G_{00} - V_2 G_{10} \qquad (27)$$
$$(E + \overline{\alpha}) G_{-\overline{1}0} = - V_1 G_{-10} - V_2 G_{00}$$
$$\ldots\ldots \qquad \ldots\ldots$$

If we proceed with the successive substitutions of the above equations and eliminate all reference to the overhead bar sites we get finally the equivalent system of equations

$$(E + \alpha_R) G_{00} = 1 - V_R G_{-10} - V_R G_{10}$$
$$(E + \alpha_R) G_{10} = - V_R G_{00} - V_R G_{20} \qquad (28)$$
$$(E + \alpha_R) G_{20} = - V_R G_{10} - V_R G_{30}$$
$$\ldots\ldots \qquad \ldots\ldots$$

where

$$\alpha_R = \alpha - \frac{V_1^2 + V_2^2}{E + \overline{\alpha}}$$

and

$$V_R = - \frac{V_1 V_2}{E + \overline{\alpha}} \qquad (29)$$

Comparison of equations (28) and (3) shows us that we have formally reduced the original problem to that of a homoatomic chain of diagonal hamiltonian element α_R and first-neighbors coupling V_R. In the present case, however, those elements are not fixed parameters but functions of the energy and must be recalculated for each new value of E considered.

Even so, what we have accomplished was a simplification of the procedures to solve the electronic structure which gives us a powerful tool to deal with more complex polymeric chains. In fact, the limitation of this approach is concerned with the dimensionality of the system considered: while topological 1-D chains can always be solved through the renormalization procedure, two dimensional networks are beyond the reach of the technique because each reormalization step would bring in an increased number of new interactions.

We will apply now the renormalization idea to the chain bridged polymers.

3.3. Renormalization of Chain-Bridged Polymers

Consider the chain-bridged polyenes of Fig. 5. First, note that within the first-neighbors approximation the electronic structure of each polymer is insensitive to the relative orientation of consecutive rings. These polymers can then be represented schematically as in Fig. 18. Also, for each specific chain we can distinguish two types of carbon atoms accordingly to their position relative to the neighbors: there are carbon atoms "internal" to the rings and "knot" carbons which connect consecutive rings.

If we use the renormalization idea to eliminate any reference to the internal carbon atoms all of these different chains are formally

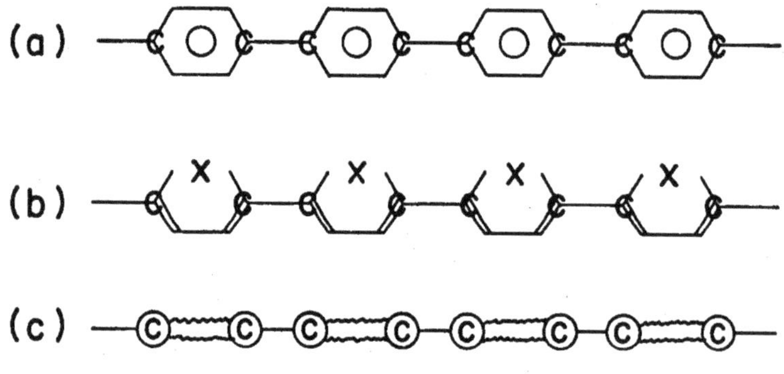

Fig.18. a) PPP and b) PPY (X=NH) and PTP (X=S) chains can be formally reduced to a renormalized PA-like chain (c).

reduced to the polyacetylene-like chain of Fig. 18c: consecutive "knot" carbon atoms interact in an alternate way through renormalized "single" and "double" bonds. Hence the problem of calculating the electronic structure of the chain-bridged polymers can be solved once and for all in a generic way through the use of T_1 and T_2 transfer functions (15) introduced for the PA case[36,37]. What we are left to do is substitute the appropriate energy functions α_R, V_1^R and V_2^R for the specific diagonal and first-neighbors interactions for each polymer.

We have seen above that to explain the transport properties of the chain-bridged polymers one had to invoke the polaron defects of Fig. 6. Although the electronic structure of these defects could be obtained in a similar way to that introduced above to treat the relaxed defect case, a much simpler procedure to handle the problem is to use the renormalization idea throughout[36,37] the process.

An interesting result comes from the analysis of the localized wavefunction associated to the defect. We have seen that the polaron can be considered as the structure which results from the attrative interaction between two solitons[17]. Since each soliton introduces a single localized energy level at the middle of the gap we have a problem formally similar to that of a hydrogen molecule: two states, one bonding and the other antibonding, must appear in the gap after the polaron creation. The corresponding wavefunctions (see Fig. 19) calculated through the renormalization technique bear this formal resemblance.

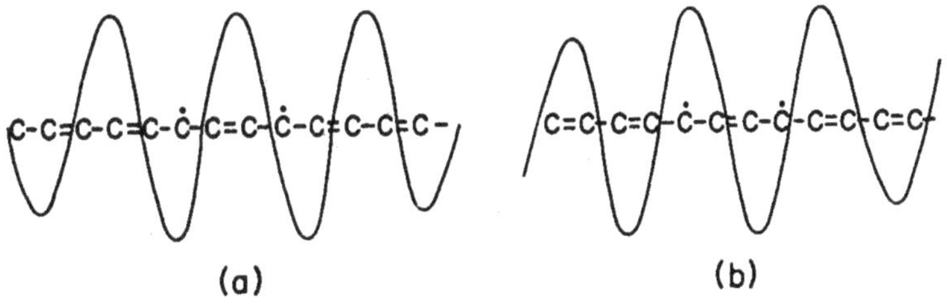

Fig.19. Wavefunctions for the antibonding (a) and bonding polaron states (b) of trans-PA calculated through the renormalization technique. Note the "nodal" plane for the antibonding case.

Let's now apply the renormalization technique to a more complex problem: the electronic structure of polyacene.

4. THE ELECTRONIC STRUCTURE OF POLYACENE

4.1. Introduction

Polyacene (PANE) is a yet to be synthesized polymer consisting of condensed aromatic rings of fused structure (Fig. 20). PANE can be viewed as two interacting PA chains and for similar reasons to those of the PA case the existence of a bond alternation pattern for the ground state of the system has been a subject of long standing interest[38,39]. More recently the high symmetry of PANE's unit cell has motivated suggestions that Peierl's instability should not occur for the polymer[40].

4.2. Electronic Structure of Regular Chains

The renormalization scheme can be applied to PANE[41] in the form suggested in Fig. 21. Note that in the equivalent renormalized chain of Fig. 21b the carbon atoms are not connected through a topological 1-D path since there is more than one way to go from one site to the next.

a)

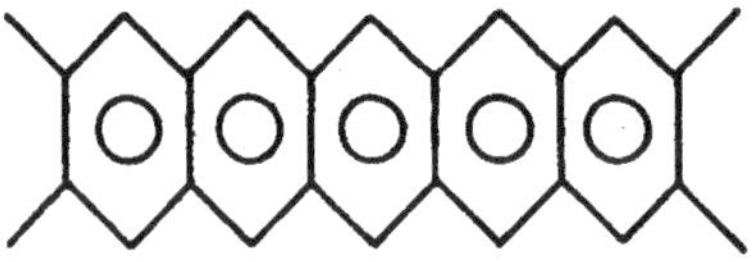

b)

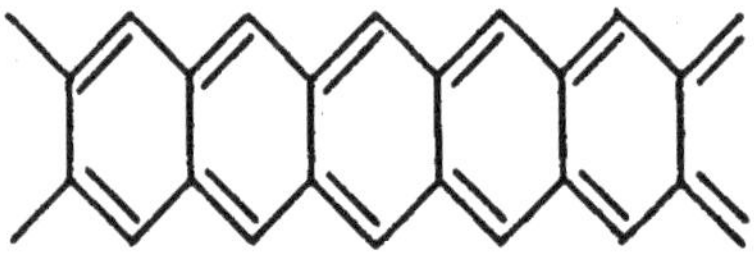

c)

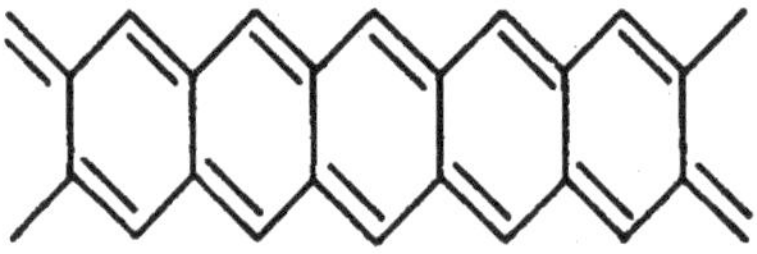

Fig.20 Possible configurations for PANE: a) nonalternant, b) alternant cis, and c) alternant trans.

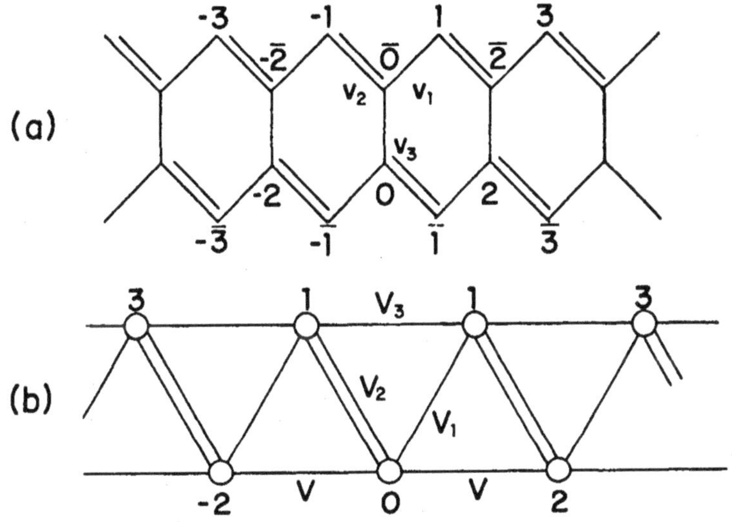

Fig.21. a) Off-diagonal coupling terms in regular polyacenic chain (for convenience the diagonal energy elements are taken equal to zero). b) Equivalent renormalized chain: note that second-neighbor interactions are introduced in the process of renormalization.

The final problem corresponds to a diatomic chain with second-neighbors interaction. After eliminating references to the intermediate sites we get the system of equations

$$(E - \alpha)\, G_{00} = 1 + V_1\, G_{10} + V_2\, G_{-10} + V_3\, (G_{20} + G_{-20}) \tag{30}$$

and for $n > 0$

$$\begin{aligned}(E - \alpha)\, G_{2n,0} &= V_1\, G_{2n+1,0} + V_2\, G_{2n-1,0} + \\ &\quad + V_3\, (G_{2n+2,0} + G_{2n-2,0}) \\ (E - \beta)\, G_{2n+1,0} &= V_1\, G_{2n,0} + V_2\, G_{2n+2,0} + \\ &\quad + V_3\, (G_{2n+3,0} + G_{2n-1,0})\end{aligned} \tag{31}$$

where

$$\alpha = \frac{v_1^2 + v_2^2 + v_3^2}{E}, \quad \beta = \frac{v_1^2 + v_2^2}{E}, \tag{32}$$

and

$$V_1 = \frac{v_1 v_2}{E}, \quad V_2 = \frac{v_2 v_3}{E}, \text{ and } V_3 = \frac{v_1 v_2}{E}. \tag{33}$$

The problem has analytic solution if we introduce the transfer matrices

$$\begin{aligned} T_1 &= \frac{G_{2n+1,0}}{G_{2n,0}} \text{ and } T_2 = \frac{G_{2n,0}}{G_{2n-1,0}} \quad \text{for } n = 0,1,2,\ldots \\ F_1 &= \frac{G_{-2n,0}}{G_{-2n+1,0}} \text{ and } F_2 = \frac{G_{-2n-1,0}}{G_{-2n,0}} \quad \text{for } n = 0,1,2,\ldots \end{aligned} \tag{34}$$

The physics of the problem is controlled by the parameters $a=V_2/V_1$ and $b = V_3/V_1$ related to the degree of bond alternation in each lateral chain and to the coupling between the lateral chains of PANE, respectively. For diatomic chains the lds per unit cell can be calculated through

$$D(E) = -\frac{1}{\pi}\left[\text{Im } G_{00}(E) + \text{Im } G_{11}(E)\right] \tag{35}$$

and is represented in Fig. 22 in units of $\epsilon=E/V_1$. Note that while the value of a determines the existence of the central gap between valence and conduction bands, b controls the degree of superposition of the two bands inside each set of valence and conduction bands. The value of the central gap, however, depends simultaneously on the values of a and b.

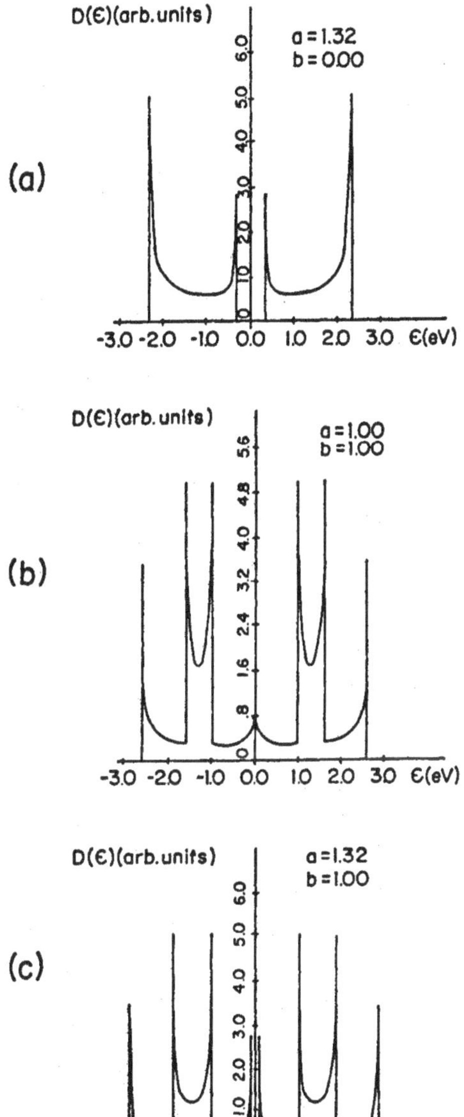

Fig.22. Lds per unit cell for polyacene for different values of parameters a and b. (a) If b → 0, the system resembles two weakly interacting PA chains. The central gap vanishes if a=1.0 (b), and is present otherwise (c).

4.3. Conformational Defects in Polyacene

One can imagine a soliton-like defect in polyacene (Fig. 23a). The corresponding electronic structure problem and the localized spin density wave function can be calculated through the use of the transfer functions (34). The local density of states for this problem is represented in Fig. 24 and the soliton wavefunction is depicted[41)] in Fig. 25.

a)

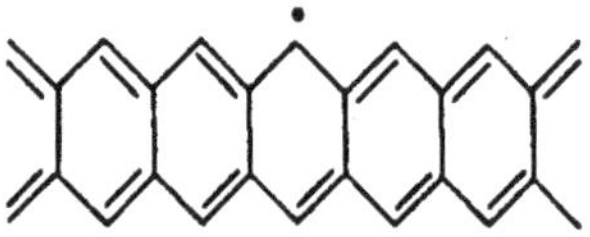

b)

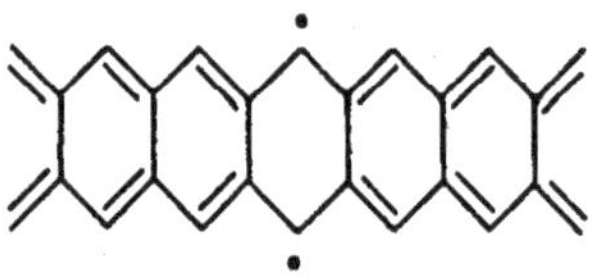

Fig.23. Possible conformational defects in PANE a) Soliton-like, and b) Polaron-like structures.

We can see from the comparison of Figures (14) and (25) that the soliton wavefunction in polyacene is identical to that of a soliton in PA. This is a rigorous result: it can be shown through the analysis of the residues of ψ_1 (ϵ=0) that the decay of the soliton wavefunction does not depend on b and is entirely determined by the bond alternation parameter a.

At this level of approximation the soliton will be confined to a single lateral chain and we can predict that the transport properties of PANE will be very similar to those of PA. However for PANE there is possibility of different kind of defects (such as the polaron-like defect of Fig.23b) and since electron delocalization is a characteristic of aromatic rings, we can also expect that inclusion of electron-electron interaction terms will allow for an interesting new physics to be associated to this polymer.

Let's discuss newt how electron-electron interaction terms could be incorporated into the Green's function treatment of the electronic structure problem.

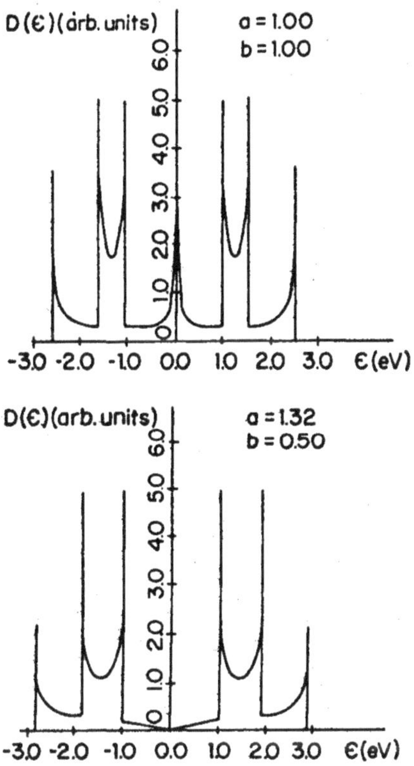

Fig.24. Lds for the central site of a soliton-like defect in polyacene for different values of a and b.

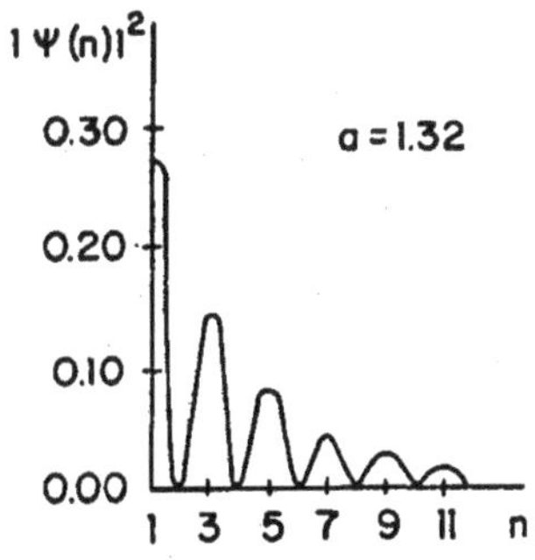

Fig.25. Soliton wavefunction for single-site defect in one of the lateral chains of polyacene.

5. EFFECT OF ELECTRON-ELECTRON INTERACTIONS ON THE ELECTRONIC STRUCTURE OF POLYMERS

As is well-known in Quantum Chemistry, electron motion is highly correlationated, i.e., change in the wavefunction of a single electron leads to immediate changes on the wavefunctions for the other N-1 electrons of a molecule[20,22]. As such, for a hamiltonian where explicit electron-electron interaction terms are included the Schroedinger equation can only be solved through self-consistent methods. In those methods an initial electron distribution is assumed and the wave equation is solved separately for each electron in the average potential of the remaining ones; a new N electron distribution can then be obtained and the process is repeated until the initial and final electron distributions at a given cycle are essentially identical.

Of course this is not the procedure in an empirical treatment such as the Huckel model we have used above, where the hamiltonian elements are taken as parameters. Good as the Huckel or the tight-binding methods may be as first approximations to the electronic structure of conjugated polymers, there are several instances in which empirical and self-consistent methods can lead to predictions which are even in qualitative difference of each other.

For instance, although the general features of the soliton wave function of Eq. (26) are observed in EPR experiments[12,42], more recent double-resonance results indicate that two different kind of sites with positive and negative spin densities are present in polyacetylene[43]. It was recognized later that inclusion of electron-electron correlation terms lead, for a soliton centered at site zero, to small negative densities at the odd sites[44,45]. The reason for this difference is easy to rationalize: although no charge transfer occur for the PA chain, the effective potential for electrons of opposite spin orientation should be different at alternate sites near the conformational defect.

On the other hand, for extended systems such as polymeric chains, use of ab-initio (literally, "from the beginning" - i.e., where no hamiltonian approximations are made) methods can become prohibitive. Although a full range of semi-empirical methods are nowadays available[20,22], there remains the question of how large a finite chain must be in order to reproduce infinite chain results (we know for instance that while observables such as bondlenghts rapidly converge with chain increase, others such as optical spectra can have less monotonic convergence)[10].

Once again the simplest way to introduce self-consistency is to use transfer function ideas. One could start with a spin-polarized version of our tight-binding hamiltonian[4]:

$$h_{i,i}^{(s)} = \alpha + J\, n_i^{(-s)}$$

where s and (-s) indicate opposite spin orientations, J is the one-center repulsion integral and n_i is the electronic spin density given by (11a) at site i. These spin densities must be recalculated at each cycle and so the correct values can only be obtained after self-consistency is attained.

We are not going to dwell on this point much further, although one should note that through simple extension the treatment suggested above can lead to very interesting results[47]: a whole set of different ground-state configurations can arise for simple self-consistent hamiltonians where one updates the diagonal and off-diagonal interactions at each cycle. The alternant polyacetylene chain can then be considered as a physical realization of one of these possibilities (a "bond-order wave").

6. CONCLUSIONS

In this series of lectures I have presented the general ideas of how techniques borrowed from condensed matter physics can be used to investigate the electronic structure of polymers. Although the examples discussed were chosen to be representative of the new area of

---

4 - This hamiltonian, introduced by Newns to describe atomic hydrogen chemisorption on transition metals, in generally known as Anderson-Newns hamiltonian.

conducting polymers, the ideas outlined above can easily be generalized to different kind of problems involving different polymeric systems (for instance, transfer functions have been used with success to study the effect of impurities in proteins[47]):

An interesting new field of investigation is related to the optical properties of conjugated polymers. Mainly due to the intrinsec anisotropy of the polymeric chain and to the characteristic polarizability of π-electron systems, conjugated polymers have been recognized as promising materials for nonlinear optics[48,49]. Different theoretical approaches have been introduced to describe the electronic structure of these systems. If, as suspected[50,51], defects also play an important role in the problem of light-interaction in these systems, transfer functions and the related techniques discussed above can be of great value to the student interested on this subject.

We should conclude, however, by repeating the word of caveat about the fact that the physics of polymers is so rich that no single approach can claim to describe the different aspects of the electronic, optical, and transport properties or these materials. To handle problems where important charge of spin disturbancies do occur, use of self-consistent methods may be indispensable. To attain the correct balance between different points of view and choose the most appropriated method to a specific problem in question is the key to progress in the understanding of the complexity and beauty of the physics of polymers.

Acknowledgements: Several people have contributed to the development of the material presented in this course. I thank in particular to Dr. H. Brandi and Dr. B. Koiller for pointing me out first the potential of the transfer functions to the study of the electronic structure of polymers. Several of the examples discussed above were done in collaboration with Dr. M.C. dos Santos, A.L. da Rosa and F. Souza. I acknowledge the continuous financial support of the Brazilian Agencies FINEP, CNPq and CAPES during the course of this work.

www.ingramcontent.com/pod-product-compliance
Lightning Source LLC
LaVergne TN
LVHW082256241026
850861LV00008B/61